Bioprocesses Including Animal Cell Culture

AKADEMIE-VERLAG BERLIN

Bioprocesses Including Animal Cell Culture

Bioprocesses Including Animal Cell Culture

Managing Editor: A. Fiechter

With 41 Figures and 12 Tables

Akademie-Verlag Berlin 1988

Die Originalausgabe erscheint im Springer-Verlag Berlin Heidelberg New York
als Volume 37 der Schriftenreihe
Advances in Biochemical Engineering/Biotechnology

Vertrieb ausschließlich für die DDR und die sozialistischen Länder Akademie-Verlag Berlin

ISBN 3-540-19004-X Springer-Verlag Berlin Heidelberg New York Tokyo
ISBN 0-387-19004-X Springer-Verlag New York Heidelberg Berlin Tokyo

ISBN 3-05-500569-4

Erschienen im Akademie-Verlag Berlin, DDR-1086 Berlin, Leipziger Straße 3—4
Lizenznummer: 202 · 100/533/88
Printed in the German Democratic Republic
LSV 1315
Bestellnummer: 763 886 1 (3070/37)
14800

Table of Contents

Process Development for Hybridoma Cells

C. Harbour, J. P. Barford and K.-S. Low
Department of Infectious Diseases and Department of Chemical Engineering, The University of Sydney, NSW 2006, Australia

This review attempts to cover those factors that would need to be considered for the optimisation of process control and development strategies for the production of monoclonal antibodies from hybridoma cell lines. The currently available experimental data for cell growth and antibody production of monoclonal antibodies is reviewed against the theoretical background developed mainly from microbial systems. The various parameters which affect the kinetics of hybridoma cell activities are then described in detail, concentrating on shear effects and oxygen requirements which are two of the most important scale-dependent effectors of cell culture growth and productivity. Finally an attempt is made to consider how all the various requirements of hybridoma cells could be incorporated into a process optimisation strategy, with particular reference to bioreactor design and mode of culture operation.

1 Introduction and Scope of Review

Mammalian cells are being increasingly employed for the production of various diagnostic and therapeutic biologicals. In particular the generation of monoclonal antibodies from hybridoma cells has attracted much attention since these reagents are now widely used in veterinary and human diagnostic assay kits, in blood grouping and in immunopurification procedures. There is also great interest in their application as immunotherapeutic agents for the treatment of infectious diseases, tumour destruction and the removal of circulating toxins and pathogens via extracorporeal shunts.

Large-scale processes are being developed in order to meet both the current and future demands for monoclonal antibodies. The current commercial requirement is being fulfilled by a number of biotechnology companies who have adopted several different strategies for bulk production.

This review will attempt to describe the kind of knowledge that is needed to establish an effective process for monoclonal antibody production and will concentrate on the requirements for a rational process development and control system with the understanding that the primary aim of the overall process is to maximise cell productivity while minimising production costs. In Sect. 2, the pertinent areas of knowledge required to achieve this aim are outlined while the current state of knowledge with regard to growth and production kinetics of animal cell culture systems, particularly monoclonal antibody production from hybridoma cells, is presented in Sect. 3.

The data is reviewed against the background of the theory of growth kinetics which has been developed to explain the growth processes of microorganisms and their product formation.

In Sect. 4 we describe the major parameters which have been identified as having significant effects on the kinetics outlined in Sect. 3. For effective process development it is important to both monitor and control the parameters, and the type of equipment which is available to do that is discussed in Sect. 5. Finally, in Sect. 5, we have attempted to describe how these various considerations can be integrated into an overall process design strategy. Two aspects are considered: 1) the type of culture method or mode of culture (i.e. batch, fed-batch, perfusion and continuous) which would achieve optimum antibody yields; 2) the techniques of cultivation or more particularly bioreactor designs which have been developed to meet the cells' biological, chemical and physical requirements. In this section, extensive reference is made to microbial growth systems (and in particular yeast growth and metabolism). This is undertaken to draw comparison and analogy with other life systems. Yeast metabolism, with its major metabolic characteristics of fermentation and respiration and the control thereof, provides an excellent analogy to the growth and metabolism of other eukaryotic cells.

2 Integrated Approach to Process Development

In cell cultures the sum of biological activities, which we see as cell growth and product formation and which involves mass transformations and kinetics, is governed by a large number of intracellular and extracellular parameters. Although the network and interdependence of the factors involved in the generation of animal cells and their products is highly complex the important parameters need to be identified. These parameters are physical, chemical and biological in nature and are not, necessarily easy to measure. It is also important to determine whether a measured variable is a parameter which actually governs the process or is a variable suitable for indicating the state of the culture. Figure 1 illustrates the major steps required for an integrated approach to product optimisation using hybridoma cell lines. Firstly, there is process simplification whereby the major metabolic variables characterising the growth of hybridoma cell lines and production of the desired product (antibody) are identified. These are then translated into a model which may be anything from something as

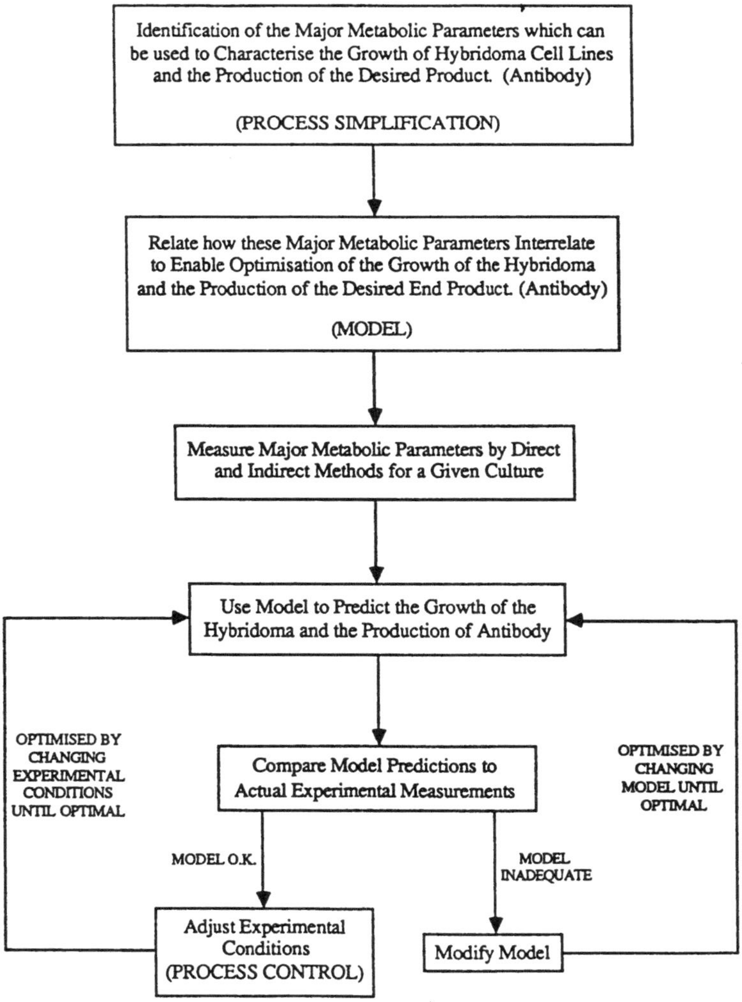

Fig. 1. Integrated approach for product optimisation using hybridoma cell lines

simple as a mass balance relating these variables to a much more sophisticated model. The decision as to what level to model the system is a complex one — a balance between simplicity, ability to measure experimentally major components of the model and the accuracy and speed of control are required amongst other things. Comparison of experimental measurements to model predictions is made (if this is possible) and an iterative process for product optimisation is commenced. This may involve model modification or process variable manipulation (or a combination of both). This process has been discussed in detail with respect to microbial growth systems [1].

Implicit from Fig. 1 is the fact that the successful optimisation of any animal cell culture system, as with microbial production systems, requires detailed knowledge of the following: 1) the pathways, kinetics and thermodynamics of cell growth and product formation; 2) the relationships of these factors to the overall cell metabolism including specific uptake rates of essential nutrients; and 3) the relationships between the cells' physiology and product formation and their external environment, i.e. the effectors of cell behaviour such as oxygen availability, concentration of essential substrates in the medium, the temperature, pH, shear stress effects and reactor design.

This kind of approach has been widely and successfully applied to many large-scale processes involving both prokaryotic and eukaryotic organisms. Computer techniques, in addition to the traditional empirical approach, are being increasingly used to construct mass and energy balances for the growth, energetics and product distribution in microbial growth processes [2]. The use of computer simulations has not yet been widely used in animal cell culture systems. There are a number of reasons for this which may be identified with respect to the integrated approach illustrated in Fig. 1. The first two steps of the process, process simplification and model formation, have not been attempted to any significant extent in hybridoma cell growth. In bacterial and eukaryotic growth, often the process may be simply reduced to a model consisting of a detailed mass balance with only a limited number of process variables required for an accurate prediction. This is often the result of homofermentative growth (only one or two major end products of anaerobic metabolism) or completely respiratory metabolism (again characterised by a limited number of end products; e.g. CO_2 and H_2O). In addition to this, the detailed pathways are well understood with respect to their stoichiometry and hence can be easily combined into a simple model (in this case, merely a mathematical form relating variables via a mass balance). Hybridoma cells have a much more complex metabolism both with respect to the range of substrates and nutrients utilised and the metabolic pathways (and hence, possible products) available. Insufficient experimental work has been undertaken to date to allow a mass balance with any degree of accuracy. Hence even the simplest model for the system is not available. The complexity of the product (antibody), where contributions to its structure are made by an extensive range of catabolic and anabolic pathways, only exacerbates this situation.

In bacterial and eukaryotic systems, simple substrate and product regimes have led to the use of a range of on-line measurement devices (notably oxygen uptake analysis, carbon dioxide production rate and ethanol excretion rate) where a direct comparison between experimental and model predictions may be made and changes to either the process variables under control or the model used to predict these variables may be made (see Fig. 1). More recently, more sophisticated on-line measuring devices (e.g. mass spectrometer and HPLC) have been used. Iterative techniques may then be used to approach product optimisation. By contrast the product spectrum of hybridoma cells does not allow on-line determination in general. Consequently while such approaches have been used extensively in microbial growth studies [3], notably in the baker's yeast industry, they have received very little attention in the field of hybridoma growth. A further consideration when making the comparison of microbial to animal cell cultivation and control is that of the time-scale of the control process. Microbial processes are generally much faster and require more rapid evaluation, comparison and control action than does animal cell cultivation. Consequently, what by necessity would be on-line in a microbial cultivation may not necessarily be so in an animal cell cultivation and the control strategy may then be considerably different.

3 Kinetics of Cell Growth and Antibody Production

3.1 Theoretical Considerations

3.1.1 Cell Growth

The growth of hybridoma cells in batch cultures (see Fig. 2) follows the classical profile observed for most microorganisms. Thus mammalian cell populations exhibit a series of growth phases:

1) lag phase — zero net growth (specific growth rate $\mu = 0$),

2) accelerating phase,

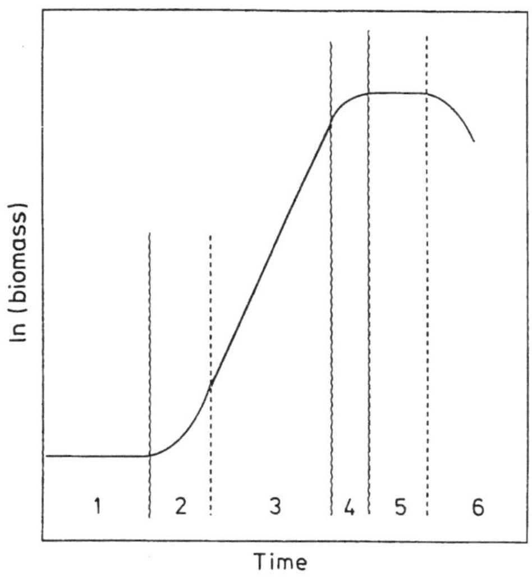

Fig. 2. Growth of hybridoma cells in batch culture

3) exponential growth phase ($\mu = \mu_{max}$),

4) decelerating phase,

5) stationary phase ($\mu = 0$),

6) decreasing phase and death.

In batch cultures environmental conditions are constantly changing; essential nutrients become depleted and metabolites and cell products accumulate. Cell mass enters the decelerating phase when either a) essential nutrients become limiting or b) inhibitors accumulate above toxic thresholds, or c) both a) and b). If the culture conditions were such that each cell was able to grow under favourable conditions, such an excess of all nutrients and the cell concentrations can be modelled by:

$$\frac{dx}{dt} = \mu_{max} x \tag{1}$$

where,

x is the biomass concentration
x_0 is the biomass concentration at zero time
t is time

μ_{max} is the maximum specific growth rate at the given conditions.

The Monod equation is applied if the growth conditions do not permit maximum growth, and where it is assumed that growth is limited by the availability of substrate, S [4],

$$\mu = \mu_{max} \frac{S}{K_s + S} \tag{2}$$

where

S is the concentration of the limiting substrate and
K_s is a saturation constant, the Michaelis-Menten constant.

Monod [5] was the first to demonstrate in 1949 that kinetics of growth, in a bacterial system, closely resembled the Michaelis-Menten equation for enzymatic substrate conversion to product in a buffer system. According to this model, cell growth was limited by the concentration of a single substrate with cells only growing at their maximum in conditions of excess limiting substrate. Since then the model has proved applicable to the growth of eukaryotic cells such as yeasts as well as prokaryotic cells. Studies concerning the continuous culture of mammalian cells have recently been reviewed by Tovey [6]. The first reports of the cultivation of animal cells in a chemically defined medium in chemostat culture were obtained with mouse LS cells growing in a chemically defined, protein-free medium [7–10], under glucose [11] and choline [12] limitation. Most of the subsequent chemostat studies, until the advent of hybridoma cells, involving animal cells have been carried out with mouse leukaemia L1210 cells [13–18] with the successful use of glucose limitation.

Stable steady-state cultures of L1210 were obtained in the chemostat at various dilution rates and the relationship between the steady-state cell concentration and dilution rate was found to be in good agreement with the theoretical curves of cell density and glucose concentration computed from the Monod equations [14, 19]. At high dilution rates however there was substantial deviation from the theoretical curves.

Recently Hu and Wang [20] have commented that these findings may not be surprising considering the complexity of mammalian cell culture media. They point out that the Monod model as expressed in Eq. (2) is for cell growth limited by a single substrate and that these growth-limiting substrates are more easily identified in chemically-defined microbial growth media than in the more complex mammalian cell culture medium containing various amino acids, vitamins, fatty acids and a serum supplement. It is therefore possible that other mathematical models [4] which describe growth independent of substrate concentration, with more than one substrate limiting, or in terms of substrate inhibition may prove more appropriate for hybridoma growth.

In a chemostat the rate of growth is determined by rate of input of fresh medium and the cells are maintained in exponential growth at a constant concentration whereby:

$$\frac{dx}{dt} = \mu x - Dx \qquad (3)$$

The specific dilution rate D relates the nutrient feed rate F to the chemostat volume V. Thus $D = F/V = 1/\gamma$, where γ is the residence time.

For steady state operation, with respect to x,

$$\frac{dx}{dt} = 0$$

and

$$\mu x = Dx$$

$$\mu = D = F/V \qquad (4)$$

Thus the specific growth rate equals the specific dilution rate at steady state. In substrate unlimited cultivation $\mu = \mu_{max}$. In his study with hybridoma cells in continuous culture Fazekas de St. Groth [21] extended the equation to incorporate a term for the significant number of dying cells.

3.1.2 Relationships Between Product Formation and Cell Growth

A rational scale-up approach requires knowledge of cellular physiology particularly cell growth and antibody production kinetics. To date there have been few detailed studies concerning antibody production kinetics reported in the literature. There are two major reasons for this:

1) The significant research commitment required to obtain accurate and reliable data.
2) Commercial sensitivity.

Kinetic data may be obtained either in batch or continuous culture (or preferably both, enabling the most complete analysis to be undertaken).

Traditionally, in microbial growth systems, batch kinetic analysis is undertaken by taking samples for biomass and product at sufficient regularity to enable a differential rate analysis to be performed. In this method over a differential time element, differential biomass and production concentrations are calculated. This primary data is then processed into the commonly quoted secondary data viz specific growth rate and specific product formation rate.

In the kinetic analysis of monoclonal antibody producing cell lines, significant differences exist from these traditional approaches in microbial growth systems, although the extent to which these differences are of significance may vary from cell line to cell line. Firstly in analysis of microbial systems, generally both the biomass and product may easily be measured accurately with high frequency. In addition both the biomass viability is high (>95%) and the product formed stable. These two

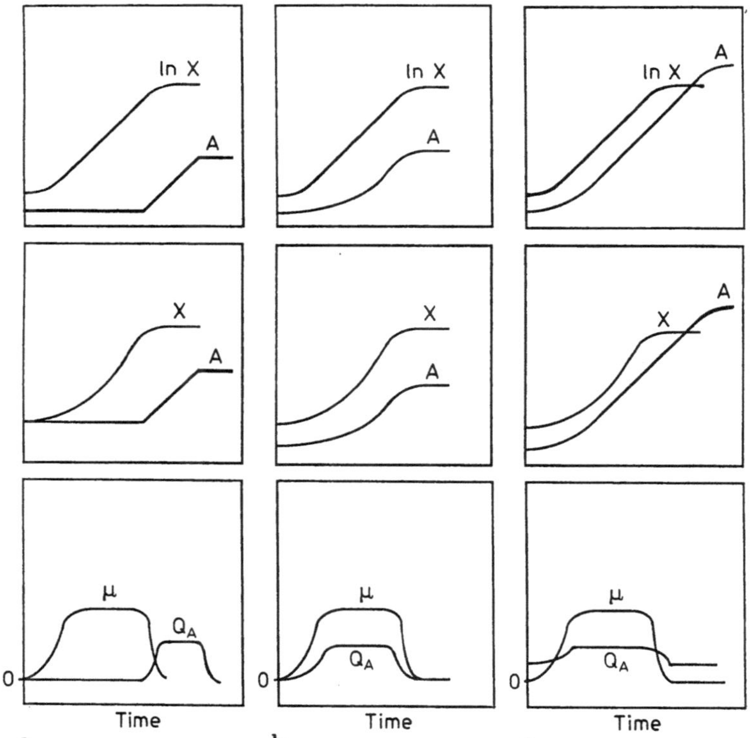

a b c

Fig. 3a–c. Typical biomass, antibody, specific growth rate and specific antibody formation rates in batch culture characteristics enable a kinetic analysis to be undertaken with high degree of accuracy and the secondary data derived from primary measurements have a correspondingly high degree of accuracy and reliability.

As a result, in microbial growth systems, typical kinetic responses have been characterised into three major types, viz growth associated product formation (see Fig. 3b), non-growth associated product formation (Fig. 3a) and combined growth and non-growth rate associated product formation (Fig. 3c). Fig. 3 illustrates these major types by taking typical primary data in two forms, namely biomass and product concentrations versus time and log biomass concentration and production concentration versus time, and illustrating their transformation into secondary data, namely specific growth rate and specific product formation rate. An underlying assumption in these profiles is that both specific growth rate and specific product formation rate become balanced within the duration of the batch.

While these major types may also be used to characterise monoclonal antibody producing cell lines, significant differences exist between these kinetic analyses and those of typical microbial growth systems. Firstly, the measurement of antibody concentrations does not have the same degree of accuracy as microbial growth products in general and their analyses are more time-consuming with the net result that they are generally taken less frequently and have less absolute accuracy. In addition the biomass may exhibit less viability than a microbial growth system. It is not clear whether non-viable cells (however measured) produce antibodies although this is unlikely and, further, whether non-viable cells release stored antibodies into the medium. Antibody degradation is also possible leading to a very complicated kinetic analysis. It may be possible to use indirect correlations for some of these estimates, e.g. the use of lactic acid concentration, although no clear evidence for the general use of such correlations exists.

Consequently it should be noted that firstly, the primary data is generally less accurate, reliable and frequent and that secondly, data derived from primary data is subject to conceptual considerations not generally associated with microbial growth systems. For example, it would generally be accepted that if viable (X_v) and non-viable (X_{nv}) biomass were present then the specific growth rate (μ) would be defined as:

$$\mu = \frac{1}{X_v}\frac{dX_v}{dt} \tag{5}$$

since, by definition, non-viable biomas is defind as biomass unable to grow and divide. However the calculation of specific antibody production rate requires the decision as to whether only viable or both viable and non-viable biomass produce antibodies. Hence if an antibody degradation rate ($K_D(k^{-1})$) is defined then the specific product formation rate may be represented in a number of forms. This basic form may typically be given as follows:

$$Q_A = \frac{1}{X_v}\frac{dA}{dt} - \frac{K_D A}{X_v} \tag{6}$$

where antibody is produced only by viable biomass

(that is $\frac{dA}{dt} = Q_A^v X_v$ = specific antibody production by viable biomass);

or

$$Q_A = \frac{1}{X_{TOT}} \frac{dA}{dt} - \frac{K_D A}{X_{TOT}} \tag{7}$$

where

$$X_{TOT} = X_v + X_{nv}$$

and antibody is produced by both viable and non-viable biomass (that is $\frac{dA}{dt} = Q_A^v \cdot X_v + Q_A^{nv} \cdot X_{nv}$ where Q_A^{nv} = specific antibody production by non-viable biomass). It is important to appreciate that these are only two possibilities based on simplistic assumptions. However, without a conceptual basis for antibody production, conversion of primary data to secondary data is not possible.

Within this context the literature on the kinetic analysis of batch data for antibody production may be discussed. Boraston et al. [22] and Velez et al. [23] have followed increases in cell numbers and antibody levels in batch cultures and their findings suggest that a considerable amount of antibody production occurs after cell growth has ceased (Fig. 3c). This data implies that a significant proportion of antibody synthesis is non-growth associated and this would appear to be the general consensus of opinion among workers in the field. There are other reports however, e.g. Lavery et al. [24], which suggest that antibody production is directly related to growth since no increase in antibody levels occurs after the peak of cell growth (Fig. 3b). These differing results may reflect the fact that each hybridoma cell line is unique with its own kinetic properties.

The kinetic characteristics of each cell line has important implications for process development and these are discussed more fully in Sect. 5. Clearly it is difficult to

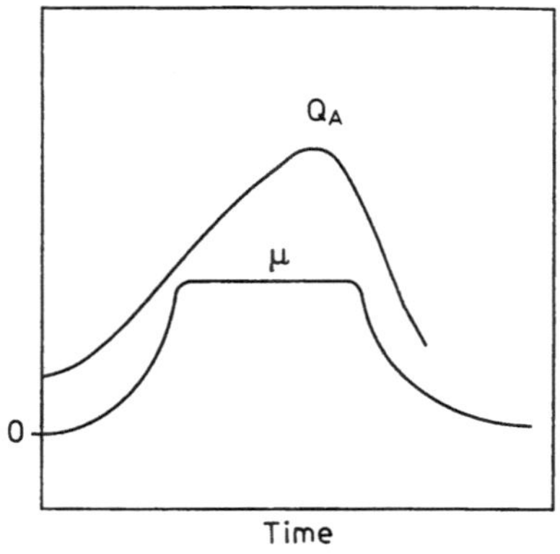

Fig. 4. Schematic diagram of experimentally obtained specific growth rate and specific antibody formation rate profiles in batch culture

determine what is happening in batch systems although Low [25] has recently compared the specific antibody production rate with specific growth in batch culture. This is shown schematically in Fig. 4. This data indicates that a proportion of antibody is produced as the specific growth rate decreases and is therefore partly not growth associated. The antibody production towards the end of the culture could be due to release of antibody from cells as they die, the antibody having been synthesised earlier in the culture.

3.2 Experimental Data

Definitive studies to determine the relationship between antibody productivity and cell growth require the use of continuous cultures. The report by Fazekas de St. Groth in 1983 [21] was one of the first concerning the continuous culture of hybridoma cells and showed the potential for the automated production of monoclonal antibodies in a cytostat. The work of Birch et al. [26] investigated the complex interaction between antibody production kinetics and cell growth. They studied the growth of one cell line in continuous culture at different growth rates with different limiting substrates, i.e. glucose, glutamine or oxygen. Their data indicated that antibody synthesis was not growth-rate dependent (see Fig. 5c) and this supported their batch culture data which showed that antibody synthesis continued during the decline phase of the culture. Also shown in Fig. 5 are three other possible specific antibody production rate trends with specific growth rate. Two of these, 5a and 5b, are illustrated for comparison purposes with Fig. 3 (namely 3b and 3c) in which growth associated antibody production (Fig. 3b) and both growth and non-growth associated antibody production (Fig. 3c) are shown. Also included (Fig. 5d) is a schematic representation of antibody production for a cell line studied in our laboratory [25]. This study indicated that antibody production was both growth and non-growth associated up to a critical specific growth rate, after which antibody

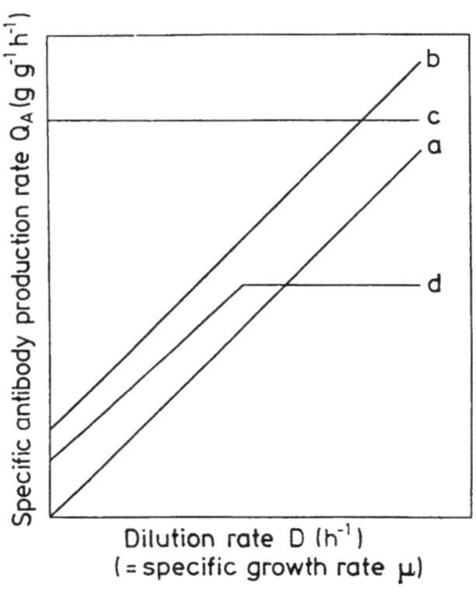

Fig. 5. Schematic diagram of typical specific antibody formation rate profiles in continuous culture

production was not growth associated. Such a response may be considered a combination of antibody production responses, illustrated by Fig. 5b and 5c. An investigation of this profile has not been undertaken in detail but the implication is that antibody production places a significant anabolic load on a hybridoma and that this reaches a maximum level at a growth rate (critical growth rate) less than the maximum specific growth rate. Consequently a saturated profile exists past this critical specific growth rate. There is evidence for such profiles in both bacterial and yeast cultivation [27].

It is then clear that significant variability in the kinetics of cell lines exists indicating substantial cell line dependence and that detailed definitive kinetic studies are few in the literature and provide a lack of a rational basis for scale-up.

4 Parameters Affecting Cell Growth and Antibody Production

Effective process development requires that the nature of all the parameters which affect the cell culture process be first identified and then measured accurately. This can be achieved by either employing the traditional empirical approach or, by adopting the more recent application of simulation experiments based on mathematical models of kinetics [2]. From the data presented in Sect. 3 the simulation approach would appear to be feasible. However, due to the complexity of mammalian cell culture systems this approach would be much more difficult than with bacterial systems. As yet little work has appeared based on this type of approach and thus most of the data presented in this Section has been obtained from an empirical approach.

In most respects the parameters which affect animal cell culture processes are the same as those identified as being important for microbial systems. However there is one important difference and that is the fact that mammalian cells, in contrast to most microorganisms, possess shear-sensitive cell membranes. This characteristic has directed those involved in the scaling-up of animal cell culture system to develop novel bioreactors which aim to reduce or avoid the use of mechanical mixing and aeration and their associated problems. Although, as discussed later, these novel processes have proved very successful, there was, until very recently, little published data upon which to base a rational system design and scale-up approach, particularly in the area of shear sensitivity.

In this review we concentrate on two of the most important scale-dependent effectors on cell culture growth and productivity, i.e. shear forces and dissolved oxygen levels. However as these problems are resolved more attention will be focused on the need for culture media suitable for the maintenance of high cell numbers. We shall therefore discuss media design as an integral part of process control and development.

4.1 Effects of Shear on Mammalian Cells

There is general agreement that the cultivation of animal cells should avoid the vigorous agitation systems employed for microbial systems. Telling and Radlett [28] found that agitation speeds used in vessels stirred by a single turbine impellor in the range 200–400 rpm were suitable for cell cultivation. Hirtenstein and Clark [29] showed that increasing stirring speed in a spinner flask caused the growth of Vero cells on microcarriers to pass through a maximum at 60 rpm. In order to explain these and many other similar observations most workers have cited the effect of shear from the agitator used to suspend the cells. In fact the word shear by itself is rather ambiguous and, although needed for proper reactor design, quantitative data about shear effects on cells are scarce. Recent reports by Cherry and Papoutsakis [30], Hu and Wang [20], Smith et al. [31] however are significant advances in this area.

In an early attempt to quantitate shear effects, Midler and Finn [32] determined the death rates of protozoa in both laminar shear fields and agitated vessels and found that extensive shear caused cell disruption. Augenstein et al. [33] used a capiallary system to investigate the shear sensitivity of mouse and human cell lines. In experiments in which cells were circulated through the capillary system at high pressure drops they found that wall shear stresses of 100 Nm^{-2} over 0.5 seconds residence time caused a significant death rate. However, it has also been shown that shear damage is a strong function of shearing time [34].

Stathopoulos and Hellmus have investigated the effects of shear on human embryonic kidney cells and shown that shear stresses of less than 0.26 Nm^{-2} caused a slight reduction in viability and no change in cell morphology. They also showed [35] that the rate of excreted cell products, i.e. urokinase from kidney cells, could also be affected by shear with maximum production occurring at a relatively low applied shear stress 0.65 Nm^{-2}. The same effect has been found on glucose consumption and lactic acid formation.

Fazekas de St. Groth [21] examined the effect of stirring speed on the growth of hybridoma cells. The growth vessel he used contained two paddles, one close to the surface of the medium and the other near the bottom. A minimum stirring rate of 10–30 rpm was required to keep the cells in suspension. Excessive agitation caused reduced growth rate with the critical agitation rate ranging from 60–100 rpm. He also found that doubling times in the stirred reactor were significantly lower than those obtained in static flask cultures which could be due to shear factors. These studies indicate that shear forces may exert more subtle forces than the blatant rupture of cells, such as inhibition of cell mitosis and the synthesis of products due to leakage of essential metabolites.

Against this background, it is important to appreciate that the experimental rationale and philosophy has been neither discussed in detail nor a general approach accepted. This may be illustrated by the following considerations.

Since hybridoma cultivation requires both aeration and agitation, the origin of shear forces on hybridomas is a combination of liquid shear and gas shear. The predominance of a particular shear stress may then have a significant dependence on the cultivation technique employed (i.e. the relative extent and individual contribution of the liquid shear from agitation and the gas shear from both agitation and aeration). Some recent work by Handa et al. [36] on the evaluation of gas-liquid interfacial effects on hybridoma viability in bubble column bioreactors have shown that, for such systems, cell viability and survival of mouse × mouse hybridomas in the presence of bubbles depended on: 1) cell type, 2) bubble size (smaller bubbles more detrimental) and bubble frequency/superficial velocity (increasing gas velocities more detrimental). While this work has begun to quantify the role of shear on hybridoma viability and antibody production, by its basis it is limited to a particular cultivation type (bubble column) and makes no reference to any contribution that mixing (other than gas mixing) may make to the total shear stress experienced by the cell. Hence, such studies make a valuable contribution to a particular aspect of the problem of quantifying the effect of shear but cannot be used in the broader sense to assist evaluating the effect of shear in different cultivation systems (where the contributions to shear from various sources may change). Thus basic study of shear stress using a range of cell lines under defined experimental conditions is warranted and of direct relevance to scale-up studies. In our opinion, it is important to quantify the likely individual contributions to the overall shear experienced by a cell to be of most use in subsequent scale-up work where the individual contributions will vary from one cultivation type and operating condition to another. Initial experiments of this kind have recently begun appearing in the scientific literature. This is an area in which both much more experimental work is required and a generalised approach is yet to be agreed upon. This area is discussed in more detail in Sect. 5.

4.2 Oxygen Requirements of Mammalian Cells

4.2.1 Oxygen Demand

Oxygen can be thought of as an essential nutrient for animal cells in culture functioning as the terminal electron acceptor in the electron transport chain and serving to provide energy via oxidative phosphorylation. It is important to identify and quantify the role of oxygen in both the catabolic and anabolic sections of cell growth. This has not been undertaken in the area of hybridoma cell growth. While it may be asserted that oxygen is predominately used for catabolic purposes (i.e. oxygen used to regenerate NADH generated from catabolic pathways), it is important to appreciate that oxygen may also play an important role in cell membrane synthesis and, consequently, the maintenance of viability. This is because of the need to synthesis components of a higher oxidation state than the substrates and nutrients from which it is made. Such considerations have not been reported in detail for hybridomas. However, given the importance of cell membranes with respect to hybridoma cell integrity (i.e. shear sensitivity, for example), this area deserves far more attention.

An important step in process control is to establish the precise oxygen requirements of the cell line used in the production system. The actual oxygen demand rates have now been determined for a variety of different cell types, i.e. Fleischaker and Sinskey, 1981 [37]; Spier and Griffiths, 1984 [38]. In a detailed study, Boraston et al. [22] determined the oxygen uptake rate of the hybridoma cell line under study to be 0.21 pmol $cell^{-1}$ h^{-1} at a maximum growth rate of 0.045 h^{-1} and that the oxygen required for maintenance purposes alone was 0.056 pmol per cell per h, i.e. approximately 30% of the total oxygen consumed by cells growing at their maximum rate is used for maintenance purposes. Similar results have been presented more recently by Katinger [39] and Low [25].

The oxygen demand of animal cells is such that due to the low solubility of oxygen in aqueous solution, approximately 0.2 mmoles L^{-1} at 37 °C [40], oxygen would be rapidly depleted from high density cell cultures unless continuously replaced. The oxygen transfer rate, OTR, (mmol or ml or μg O_2 transferred per unit volume per unit time, usually mmol O_2 L^{-1} h^{-1}) from gas phase to culture medium of any vessel can be expressed as:

$$OTR = K_La(C^* - C) \quad (8)$$

where

K_L = oxygen transfer constant of coefficient
a = the area of the interface across which oxygen transfer occurs
C^* = the amount of oxygen dissolved under saturating conditions per unit volume
C = the amount of oxygen dissolved per unit volume measured at any time during the reoxygenation

The provision of an adequate oxygen supply to mammalian cells is a critical factor which has to be considered for successful scale-up. Since oxygen supply introduces a gaseous phase into a liquid phase, any effect resulting such as shear phase on the cultivated cells needs also to be considered when aeration strategies are developed. Glacken et al.[41] have stated that oxygen limitation for a culture of HeLa cells at 10^7 ml^{-1} (respiring at 5.0 mmol O_2 L^{-1} h^{-1}) would occur at a volume of less than 1.0 L when the head-space of the vessel is filled with air. If oxygen replaced air in the head-space then the volume before oxygen limitation occurs could be 3.5 L. Clearly this means that for larger culture volumes oxygen must be directly introduced into the medium. The sparging of air or oxygen directly into the liquid medium has been used in the cultivation of suspension cells[42] but is not common practice due to foaming in serum-containing medium and the potential shear effects reviewed in the previous Section. Thus the supply of adequate oxygen for animal cells within a low-shear environment has been the major aim of those involved in large-scale process development. This type of work has led to the development of several alternative strategies which will be reviewed later.

4.2.2 Optimum Oxygen Levels for Cell Cultivation

The dissolved oxygen level has significant effects on cell metabolism[38,43–45] influencing both cell growth and the generation of cell products. Spier and Griffiths[38] have reviewed the optimum concentration required for growth by a number of different cell lines. It is clearly important for effective process control to know whether or not the optimum oxygen concentration required for growth is the same as that for maximum productivity. Mizrahi[45] found that both cell growth and immunoglobulin production were oxygen dependent. However, whereas the highest cell yield was obtained at 100% air saturation (approximately 160 mm Hg pO_2) the maximum immunoglobulin yield was achieved at 33 mm Hg pO_2. Boraston et al.[22] reported that the cell growth and antibody production of a murine cell line was unaffected over the range 8–100% of air saturation. In contrast, Reuveny et al.[46,47] showed using a murine hybridoma cell line that, although cell growth was optimal at an oxygen level of 60% air saturation, more antibody was produced during an 8-day batch culture when the oxygen level was maintained at 25%. These results clearly indicate that each cell line has to be fully characterised with regard to its physicochemical requirements prior to process scale-up, although it is recognised that this may not be commercially feasible.

4.3 Media

The recent development of bioreactors which permit the cultivation of cells at high densities, has focused attention on media design and supply. Most media used for the cultivation of mammalian cells were developed for batch cultures of relatively low density cell populations and are therefore generally not optimal for bulk production of biologicals including monoclonal antibodies. Medium design for culture should also take into account downstream processing for product recovery since this is a major cost factor on the production of biologicals from mammalian cells. Thus the development of an appropriate medium for a particular cell line should not be carried out in isolation from the overall process design and control planning. The validity of this approach is now being recognised and in an excellent recent review Hu and Wang [20] have discussed medium design from a kinetic point of view. They point out that the optimisation of mammalian cell culture medium includes the following: 1) manipulation of medium composition, including the concentrations of serum and/or growth factors; 2) the balancing of the rates of supply and consumption of nutrient; 3) the removal of inhibitory metabolites while maintaining the appropriate levels of growth-stimulating metabolites. They state that because of the complexity of mammalian cell culture media it could be virtually impossible to examine all possible combinations by trial and error to define the optimal conditions. Thus it is necessary to develop kinetic models to fully characterise the operating parameters and then test the models experimentally. There are as yet few reports relating to the application of this type of approach to process design and control for the production of biologicals from mammalian cells. There is in contrast, much published data concerning microorganisms, particularly bacteria and yeast, which could prove useful to animal cell technologists. An often unconsidered aspect of media design is that of its effect on waste treatment strategies. Many examples exist in this area of microbial growth where the media used has a direct (if not dominating) effect on waste treatment stages. For example, in the molasses based substrates, the traditional processing in the sugar industry results in a high level of sulphate in the substrate and, consequently, the waste. This has a significant effect on on-line waste treatment strategies employed following the fermentation process. Large-scale hybridoma cultivation will undoubtedly provide a waste treatment problem and the specific effect of various media components should be part of any strategy for media development.

In order to carry out a systematic approach to media design, chemostat methods are the most effective mechanism as has been shown by Mateles and Battat [48]. In a more recent review Goldberg and Er-el [49] described improvements in the methodology of medium design, particularly the usefulness of pulsing methods in chemostat experiment. Fiechter [50] has also shown the useful type of information that can be obtained by pulsing a potential nutrient or growth factor into a culture during steady-state. Whereas the non-specific reactions to an addition indicate the non-essential nature of the nutrient, an exponential increase following addition indicates a definite limitation of the nutrient. Quantitative as well as qualitative information can be obtained thus determining the required amounts of essential nutrients. Recently, advances in experimental methods available for media analysis has assisted such a structured approach. Inductively coupled plasmid emission spectrophotometry, for example, allows the analysis of a wide range of anionic and cationic media components rapidly and with high accuracy [51]. Such techniques can only assist in the quantitative analysis of medium requirements.

In his review describing the physical and chemical parameters affecting microbial growth, Fiechter [50] has outlined a step-wise approach to the systematic design of media involving the selection of essential components, preparation of the medium and the preliminary plots of biomass (X) versus dilution rates (D) in chemostat cultures. In the preliminary work, plots of biomass and limiting substrates (S) are carried out in order to determine the growth constants (Y, k_s, μ_{max}) and to determine whether or not the growth kinetics fit the Monod model. The utilisation rates of substrate and oxygen, CO_2 production and respiratory quotient are also computed along with the mass balance. The carbon-limited medium is then tested by chemostat pulse/shift experiments for possible hidden limitations that were not detected in batch experiments to determine carbon limitation. Finally plots of biomass versus dilution rates are performed and then a complete diagram for x, s, and specific production rate (qp) is constructed in function of D. The q_s, q_{O_2}, q_{CO_2}, q_p and RQ are calculated and the values obtained for Y, k_s, and μ_{max} compared with the values known for the strain. It is doubtful whether this elegant method is suitable for medium improvement for hybridoma cells on a routine basis.

This approach has worked well for the proper identification of the parameters influencing microbial growth.

The complexity of mammalian cell culture media is such that for the majority of cell lines the major growth limiting parameters have not yet been identified conclusively. The following factors could all play a role: unfavourable pH; the depletion of essential nutrients such as glucose and/or glutamine; the accumulation of inhibitory metabolites, e.g. lactate and/or ammonia; media components contributing to membrane structure and integrity. Detailed analysis of hybridoma cell lines in this context is in its infancy. In the following Sections we examine these factors in more detail.

4.3.1 Limiting Nutrients

Glucose is present in relatively large concentrations (5–25 mM) in most cell culture media and is consumed most rapidly among the components of the medium, although it is not clear whether it is the only growth limiting substrate in all circumstances. Glutamine is also an important substrate and has been recognised by Reitzer et al. [52] as an alternative carbon and energy source for HeLa cells. Zielke et al. [53] have provided evidence that the glutamine uptake rate is affected by glucose concentration in some cell culture systems and thus the two are discussed together. Reuveny et al. [46] have also reported on the interdependence of glucose and glutamine utilisation by hybridoma cells.

As described previously, the systematic quantitative determination of media requirements is essential for effective process control and involves the determination of substrate uptake rates and growth yields. Low and Harbour [54, 55] described the kinetics of glucose utilisation by two different murine hybridoma cell lines and calculated the glucose quotients and approximate yields for the two cell lines in static flask cultures. Their data showed that the glucose quotients (mmol g^{-1} cells per h) were independent of initial glucose concentrations. In contrast Thomas [56] has shown that for most cell lines the concentration of glucose in the culture medium is critical for determining its rate of utilisation. Low and Harbour [55] compared the cell yields obtained per mole of glucose with the two cell lines with those obtained by other workers using different cell lines. The hybridoma cell lines utilised glucose more efficiently than MRC-5 [57] cells but less so than BHK [58] and LS cells [9]. They also showed [55] that the initial glucose level in low-glucose DME, i.e. 5 mM, limited final cell yields. When the initial glucose level was increased to 11 mM both cell and antibody yields were significantly increased although further increases in glucose concentration had no effect, suggesting either pH inhibition due to lactate production or that other substrates became limiting at the increased cell density. Reuveny et al. [46] in an excellent paper concerning the factors affecting cell growth and monoclonal antibody production in stirred reactors have shown that after 2–3 days of batch culture both glucose and glutamine become limiting factors for hybridoma cell growth and antibody production. They also showed that glucose uptake rates were affected by dissolved oxygen levels; that is, at 25% and 60% air saturation in spinner flasks the glucose uptake rates were respectively 0.9 and 15 mg $\times 10^{-6}$ cells per day. These results are in contrast to those presented by Boraston et al. [22] who reported no effect on the rate of glucose utilisation in culture of hybridoma cells maintained at dissolved oxygen levels in the range 8–100%. Reuveny et al. also showed that supplementation with glucose or glutamine to avoid limitation led to an increase in cell numbers between days 2 and 4 but then cells died as rapidly as controls. However the combined feeding of glucose and glutamine significantly contributed to increases in both cell yields and antibody production; that is, 295 µg ml^{-1} of monoclonal antibody were detected in the glucose-glutamine fed culture compared to 150 µg ml^{-1} in the control, 160 µg ml^{-1} in the glucose-fed and 170 µg ml^{-1} in the glutamine-fed cultures. Clearly then attempts to regulate glucose and glutamine supply are important for optimising antibody yields. It is also important to consider the production of the potentially inhibitory metabolic wastes derived from glucose and glutamine, i.e. lactate and ammonia.

4.3.2 Metabolic Inhibitors

Lactic acid is generated in cell culture systems mainly from glucose metabolism. Reuveny et al. [46], found levels of up to 2.3 mg ml^{-1} after cell growth in spinner cultures and these levels were dependent on dissolved oxygen levels; that is, where dissolved oxygen was controlled at 75%, 60%, and 25% of air saturation the rates of lactic acid production reached 1.2, 0.7, and 0.3 mg $\times 10^{-6}$ cells per day respectively. They calculated that approximately 17–23% of the glucose was converted to lactic acid. Boraston et al. [22] reported even higher levels of lactate accumulation, i.e. 3.35 mg ml^{-1}, in an airlift culture of a murine hybridoma and, in contrast to the work of Reuveny et al. [46] showed that dissolved oxygen levels had no effect on lactate accumulation over the range 8–100% of air saturation. Low and Harbour [55] have also reported high levels of lactate production by hybridoma cell lines and from the lactate and glucose quotients concluded that there was an almost 100% conversion of glucose to lactate.

The levels of lactate accumulation reported in hybridoma cell cultures could exceed the buffering capacity of the medium resulting in pH values lower than optimal. Reuveny et al. [46] have investigated the proposition that lactate is growth inhibitory by adding lactic acid in the range 0.5–2.5 mg ml^{-1} to hybridoma cell culture shake flasks. They found that lactic acid levels above 2.5 mg ml^{-1} acid did exhibit toxic effects manifested as depressed cell and antibody yields, although these effects may be reduced with adaptation.

There is a growing body of knowledge concerning the amino acid requirements of animal cells [58–62] although as yet little dealing specifically with the requirements of hybridoma cells in culture. Griffiths and Pirt [63] showed that rates of amino acid consumption vary with the specific growth rate. Most studies show that glutamine is the most rapidly utilised amino acid and although some glutamine is essential for anabolic processes in the cell, its major function in some cell lines, e.g. HeLa cells, is in aerobic energy metabolism in which deamination results in production of ammonia [52,53]. Glacken et al. [41] have reported that prolonged exposure to more than 4 mmol L^{-1} of ammonia can inhibit cell growth and that 1 L of culture of HeLa cells growing from a cell density of 3.4×10^5 cells per ml to 1×10^7 cells per ml may produce up to 30 mmol of ammonia per L. Butler [60], working with microcarrier cultures of MDCK cells, reported that ammonia levels of between 2–3 mM were toxic to cells. Others [64,65] have presented similar data for other cell lines. Reuveny et al. [46] monitored the levels of ammonia accumulation during batch cultures of hybridoma cells growing in 100 ml spinner flasks and observed toxic affects on cell growth at concentrations above 2 mM when ammonium chloride in the range 1–5 mM was added to growing cells. They calculated that during the later stages of cell growth in batch culture, when they detected 4.5–5.5 mM ammonia, toxic effects would cause a decrease in cell viability and antibody production.

Several strategies have been developed to overcome the problem of growth inhibition by lactate and ammonia including perfusion of media which is discussed later. One approach has been to modify the culture medium so that the cell's metabolism is altered to prevent accumulation of inhibitory waste products. In 1958, Eagle et al. [66] substituted galactose for glucose as the carbohydrate source in the medium, resulting in a 67-fold decrease in lactic acid generation. Fleischaker [67] reported a similar effect when glucose concentration in the medium was kept relatively low, i.e. 0.5 mM vs 20 mM, by supplying glucose to FS-4 cells in a fed-batch fashion. The sophisticated process control of glucose metabolism used by this author is discussed later. Glacken et al. [41] described how ammonia production can be reduced to less than 40% by continually feeding glutamine to the culture, thereby keeping a steady, low concentration of glutamine, i.e. 0.2 mM vs 4 mM. In other attempts to replace glucose, Cristofalo and Kritchevsky [68] cultured WI-38 cells in media containing glucose, mannose, fructose or galactose and found that these carbohydrates could substitute for glucose. These substrates were utilised less rapidly than glucose, particularly fructose, and less lactate was produced per mole of fructose and galactose than per mole of glucose and mannose. Imamura et al. [69] found that when MDCK cells were grown in media containing either glucose, fructose or maltose, cell yields were similar although less lactate was produced from fructose than from glucose. An analogous situation occurs when yeasts are grown on a number of different carbon sources. Generally, carbon sources other than glucose result in a lower growth rate and consequently a more respiratory growth (since the respiratory capacity of yeasts is limited), and fewer intermediates (e.g. ethanol) being produced. In yeast physiology, considerable debate still exists regarding the role of specific sugar substrates (particularly glucose) and specific growth rate and their interaction. Even though this is not completely understood in yeasts, considerably more experiment data exists on which to form a theory (however, such theories are still contentious). Animal cells such as hybridoma cells have not been subject to the extensive experimentation of yeasts.

Reitzer et al. [52] grew HeLa cells on either glucose, galactose or fructose and although growth was similar with each substrate the metabolic pathways were found to be different. When glucose was present at >1 mM concentration, 80% was converted to lactic acid and only 4–5% entered the TCA cycle. In contrast, when the cells were growing on fructose the glycolytic pathway was almost inactive with most of the fructose carbon entering the pentose-phosphate pathway. Thus the replacement of glucose with fructose in order to culture cells which produce reduced levels of lactate from fructose would facilitate pH control. Low and Harbour [55] used this approach with two hybridoma cell lines with mixed results. They reported that one cell line produced significantly less lactate when grown on fructose compared to glucose, whereas the other cell line was able to metabolise fructose as rapidly as glucose and as a consequence produced equivalent amounts of lactate. Reuveny et al. [46] tried to prolong cell viability of hybridoma cell cultures by substituting either galactose, fructose, maltose or starch for glucose in the medium. They found that although all four of the replacement carbohydrates were metabolised by the cells, none increased cell yields or prolonged viability compared to glucose. In fact, apart from maltose which produced similar results to glucose, the other substrates decreased cell yields by 25% and antibody concentrations by 30%. These results clearly indicate that each cell line must be fully characterised metabolically as part of the process control and development process.

4.3.3 Defined Media

The content of the previous Section has illustrated both the complexity of mammalian cell's culture requirements and the complexity of the media in which they are cultivated. Thus a detailed analysis of the nutrient needs of hybridoma cells requires that they first be cultured in a completely defined, i.e. serum-free, media although this approach may not be commercially viable or even necessary to achieve maximum productivity. Reduction or elimination of serum from the medium has the added advantage of reducing protein load thus facilitating product recovery during downstream processing. There is also the need to satisfy licencing authorities who might require that a particular mammalian cell-derived biological be produced in a serum-free environment. Cost reduction is another driving force in the development of serum-free media although paradoxically the currently commercially available serum-free media are significantly more expensive on a laboratory-scale than conventional media with a 10% foetal calf serum supplement. These costs may not reflect the cost of large-scale hybridoma production. Despite the high initial costs of serum-free media it is possible that this could be offset by significant savings in downstream processing due to the reduced protein load in the medium. Reduced serum levels will also avoid the foaming problems associated with culture in air-lift and stirred bioreactors, although the reduced protein load could lead to increased shear damage in stirred reactors, requiring the addition of protective agents [70, 71] some of which, increase media viscosity.

There are now a number of reports and reviews describing the cultivation of various mammalian cells including hybridoma cells in serum-free media [21, 72–81] and this subject will not be covered in detail here. In situations where serum-free media are inappropriate, it could be advantageous to determine optimal foetal calf serum levels using relatively simple experiments relating cell and antibody yields to different initial foetal calf serum levels. For example, Low and Harbour [82] reported the effects of foetal calf serum levels on different cell lines growing in different media. Reduction of serum levels from 10% to 5% did not affect antibody yields in any of the media tested, whereas at 2% serum levels yields were significantly lower. Thus significant cost savings were achieved simply by halving the required serum supplement. In addition, the cells appeared to be quite stable over prolonged periods of culture at limiting serum levels [82]. The authors also showed the suitability of a commercially available serum-free medium.

As stated there are now a number of reviews in this area and it is clear that, although it is practically feasible to produce monoclonal antibodies in a serum-free environment, it may not be commercially viable due to the increased media costs and the possibility of decreased cell and antibody yields due to shear damage or other effects in media which are not fully optimised.

4.4 Temperature

Temperature is clearly an important variable to monitor and control in cell culture although little data has appeared concerning the effects of temperature on hybridoma cell growth and antibody productivity. It has been shown that cultured animal cells remain viable for long periods of time under lower than optimal temperature [83]. Reuveny et al. [46] studied the effects of four different temperatures on antibody production and showed that although cells remained viable for longer periods at temperatures less than 37 °C, the antibody yields were significantly decreased. This was explained by the fact that the rate of cell metabolism under low temperature was reduced significantly as measured by the rate of glucose consumption. In contrast, Himmler et al. [84] reported that the specific monoclonal antibody production rate of agarose-immobilised hybridoma cells was the same at 32 °C and 37 °C.

When considering temperature effects on the growth of cells, it is important that certain factors be taken into account. Firstly, it is essential to understand the temperature region in which one is operating with respect to the sub-optimal, optimal and super-optimal temperature tolerated by the cell [85]. This is particularly important when temperature changes occur during any process. Secondly, it must be appreciated that the effect of temperature on the catabolic section of the cell may (and often is) considerably different [86]. Consequently, a clear understanding of these separate effects and any interaction is essential. Whilst this has been undertaken in microbial processes, such as cheesemaking [86], no experimental data has been published with respect to hybridoma cells.

4.5 pH

It is important to establish optimal pH range for growth of cells since this may vary considerably from cell line to cell line.
pH affects cell survival, attachment, growth and function.

Therefore, in order to optimise culture conditions optimal pH has to be maintained. Thomas [56] has recently reviewed the pH requirements of cell culture systems and it will not be dealt with in detail in this review. The pH range of 7.2 to 7.4 is the most commonly used in mammalian cell culture, but some cells can survive very well in the pH range of 6.6 to 7.8 [87–89]. Most hybridoma cells are cultured in the range of 6.9 to 7.4.

Eagle [88] showed that normal human and rodent fibroblasts tended to have a more alkaline pH optimum than transformed cells from the same species. Barton [89] found that HeLa cells would achieve similar final cell densities in the pH range of 6.8 to 7.5 but the doubling time of these cells became shorter when the pH was above 7.2. In batch and feed-batch cultures, the pH of the culture medium can change significantly during the course of a culture cycle. Ceccarini and Eagle [90] found that wide fluctuations in culture pH tended to affect final cell yields. Birch and Edwards [91] found that pH affected the growth rate and population density of human lymphoblastoid cells and that in different phases, pH affected the cell's efficiency of glucose utilisation but not the conversion of glucose to lactate.

A pH which is optimal for cell growth may not be optimal for product formation or activity and, as yet, there is little data available in the literature concerning the effect of pH on monoclonal antibody yields. In considering the effect of pH, it is important to differentiate bet veen the effect of pH *per se* and the effect of the products responsible for the pH effect. In microbial growth, it is clear that such differentiation is of considerable importance. Anaerobic digestion systems, for example, are generally thought to be affected with respect to stability by the production of volatile fatty acids. Whether this effect is a direct result of the acidic product or a direct result of the effect of pH *per se* has only recently been examined. It would be generally accepted that the acidic products and not the pH *per se* are responsible for anaerobic digestion instability. Such an analysis has not been undertaken for hybridoma growth. A similar effect has been found with lactic acid bacteria in which the importance of differentiating between pH *per se* and lactic acid has had a considerable effect on the understanding of the kinetics of growth of lactic acid bacteria.

There are several ways of controlling pH. The most commonly used are the buffer systems involving CO_2/HCO_3^-, HEPES, acid/base, or air/CO_2 combinations. Alternatively, pH can be indirectly controlled by attempting to restrict the amount of lactate produced by the cells. This method which involves the use of alternative sugars to glucose is discussed in more detail in Sect. 4.3.2.

In using the CO_2/HCO_3^- buffer systems, it is important to bear in mind that this system has a low capacity in maintaining medium pH above neutrality when excess amount of lactate is produced by the cells. This is commonly observed in batch culture runs in sealed vessels. For cells which produce large amounts of lactate, a medium containing high concentrations of HCO_3^- (e.g. 3.7 g L^{-1} in DMEM) will serve as a better buffer than one containing low concentrations of HCO_3^- (e.g.

Ham's F12 or RPMI 1640). These latter media are better suited to cell lines which produce large amounts of CO_2.

The addition of HEPES in combination with HCO_3^- to a culture medium produces a buffer with higher buffering capacity (e.g. Iscove's modified Dulbecco's medium). Low [25] has shown that HEPES, by itself, at 25 mM concentration could not elevate the pH of a commonly used medium (e.g. DMEM without $NaHCO_3$) above pH 6.7 and, as a result, very low hybridoma cell yields (2×10^4 per ml after 6 days incubation) were achieved in this medium. Increasing the concentration of HEPES, however, would have the effect of raising the osmolarity of the medium to a level unfavourable for optimal cell growth.

An acid/base system (e.g. NaOH/HCl), when used in excess amount, has the tendency of increasing medium osmolarity beyond the optimal level. Low [25] has shown that by the addition of 0.1 N NaOH to control medium pH, the osmolarity of the growth medium was raised by 48 m Osm per L after 6 days of incubation. Clearly, this is a system to avoid when maintenance of an optimal culture medium osmolarity is desired.

Hybridomas, like other transformed cells, use glucose inefficiently. For example, Low and Harbour [55] reported a conversion of glucose to lactate of almost 100%, resulting in a fall in medium pH. Therefore, one way of avoiding large pH changes would be to reduce lactate production. Zielke et al. [53] optimised culture conditions for human fibroblast cells by maintaining very low levels of glucose in the medium through daily feeding of glucose to the culture.

Taylor et al. [92] increased the O_2 tension of mammalian fibroblast culture to achieve a reduction of lactate production. Similar results were also reported by Kilburn et al. [93]. However, Boraston et al. [22] found that the rate of lactate production by hybridoma cells was not affected by dissolved oxygen level.

One other method of controlling pH is to use a carbohydrate other than glucose in the culture medium, and this is discussed in more detail in Sect. 4.3.2. However, Low and Harbour [55] found that some hybridoma cells can metabolise fructose as efficiently as glucose, resulting in the production of high levels of lactate.

5 Process Optimisation Strategies

The major aim of process development for the production of monoclonal antibodies from hybridoma cells is clearly to maximise functional antibody yields while minimising production costs. The demand for antibodies is such that they need to be produced on a large scale and so high product yields must be maintained from laboratory through pilot to final production phase. There are a number of ways of achieving this aim including:

1) increasing the expression levels of the cells, e.g. by using those hybridoma cell lines with high secretion levels or by attempting to improve existing cell lines with rDNA techniques. Although this approach is not within the scope of this review for hybridoma technology, it is very important in the process development of other rDNA biologicals produced from mammalian cells. There have also been reports of improvements to antibody secreting cell lines, particularly lymphoblastoid cell lines which generally have low secretion rates and poor stability [94–96].

2) Increase the cell culture volume of the process;
3) increase the cell concentration within the culture.
Areas 2) and 3) will be discussed as part of this Section dealing with the strategies which have been adopted for optimisation of antibody production. It is of fundamental importance that in order to achieve the stated aim of maximising antibody production, a rational process control and development process is developed. At its most basic level this would involve a characterisation of the cell line to be used in the process, i.e. test runs to determine whether or not the cells can be grown to industrial scale; an assessment of cell stability, level of productivity and secretion rate; checks for contamination with microbial agents; and attempts to cultivate cells in inexpensive media. At a more sophisticated level an appropriate biological process control system involves a network of activities, as was shown in Fig. 1.

For this type of approach to work in practice requires the acquisition of a good deal of basic data such as: 1) a detailed description of cell growth patterns and kinetics including an estimation of the growth constants (y, K_s, μ_{max}); 2) the pathways, kinetics and thermodynamics of the product formation and the relationships to overall cell metabolism, including the specific uptake rates of essential nutrients; 3) the relationship between the cell's physiology and product formation and various physical and chemical effectors, e.g. shear stress, temperature, pH, oxygen levels, glucose and glutamine concentrations. Although few publications have appeared in the literature dealing specifically with hybridoma cells and monoclonal antibody production in these terms, we have attempted so far in this review to describe relevant information which could form the basis for this type of approach.

Such extensive experimental data requires some overall synthesis to enable an overall view of cell metabolism and product formation to be achieved. Mathematical models and computer simulations have traditionally (in the area of microbial growth) provided this link. Such approaches have not been reported in the area of hybridoma cell growth.

In addition to the acquisition of the basic kinetic data, it is also necessary to have sophisticated hardware and software to measure, monitor and control the essential parameters in the production process.

5.1 Measurement and Control of Parameters

Effective instrumentation is required to accurately measure and control the key parameters affecting growth and production kinetics that have been discussed previously in this review. Also when cells are cultured on a large scale it is important that each run should be reproducible and that cell and antibody yields be maintained at the optimal level. In 1983, Glacken et al. [41] commented that most mammalian cell production units described in the literature at that time did not utilise process control. They stated that the lack of sophisticated methodology for the process control of mammalian cell culture was not a problem of scale but due to a failure to develop process control strategies to optimise cellular metabolism. This situation is now changing rapidly and more recently Fleischaker [67] has described a more sophisticated appropriate process control strategy for mammalian cells (Sect. 5.2).

The range of instruments and sensors available and suitable for use in mammalian cell cultures has been adequately reviewed by Fleischaker [67] and will not be discussed in detail here. It is important to stress that the various sensors need to be extremely reliable since the production processes including mammalian cells are much longer than those including microorganisms. For example, batch production runs may last for 4 up to 10 d, perfusion cultures could extend for more than 20 d and continuous cultures for over 30 d. In addition, mammalian cells are more sensitive than microorganisms to trace chemicals which may leak from sensors, e.g. the leakage of lead from oxygen electrodes [25].

Harris and Spier [97] have recently comprehensively reviewed the measurement and control of chemical and physical parameters in process systems concentrating on those conditions which prevail in the unit process cultivation of animal cells. In this excellent review the authors have described in detail the various probes available for use in mammalian cell culture systems. In another review, Harris and Spier [98] have described the application of computers in animal cell biotechnology for process control. As an example they describe how computation of the stirrer speed or gassing rate can be used to gauge the culture's oxygen demand and hence determine the end of the growth phase or any change in steady state conditions. In addition, parameters, such as the respiratory quotient, may be passed to digital control loops and controlled in the same way as directly measured parameters, providing the possibility of directly controlling biochemical factors rather than working indirectly via pH and dissolved oxygen. Thus the process control of animal cell culture systems is becoming more refined although as yet little detailed information has appeared in the literature relating specifically to control of hybridoma cell processes. It is likely that most of this type of work has been carried out as part of commercial research and development and thus remains proprietary information. An exception is the work of van Wezel et al. [99] who have employed sophisticated process control to optimise antibody production yields. This group has developed the so called Bilthoven Unit which has been used both for cells growing in free suspension and in microcarrier culture. The system consists of a control panel with a conventional electronic monitoring and control unit to which bioreactors in the range 3 to 500 L can be connected for control of temperature, pH and dissolved oxygen. In order to optimise cell growth and product generation the control system has been extended to include other parameters such as CO_2 and glucose concentration. Using this control system, on-line (temperature, pH and dissolved oxygen) and off-line (glucose, NAD, etc.) signals can be measured and controlled.

In addition to instrumentation and computer control, the actual cultivation method is an important factor to consider in the production process.

5.2 Mode of Culture Operation

In this Section and the next where we consider bioreactor design, we are primarily concerned with strategies for maintaining dense cell population expressing product at maximum rate. We have already discussed the optimisation of media for cell growth and shown that in some cases sufficient data is now available concerning yield coefficients to be able to adjust the substrate concentration to generate a specified cell and hence antibody yield. The optimised medium then has to be supplied to the cells at the appropriate time and speed to maintain maximum productivity and this will depend on the choice of cultivation method, i.e. the choice between batch, fed-batch, perfusion or continuous modes of operation or a mixture of these methods.

Batch cultures would appear to be the least suitable system for obtaining maximum cell and antibody yields. During these cultures dramatic changes occur in the cells' chemical environment with the decrease of essential nutrients and the accumulation of toxic metabolites occurring simultaneously. There is also some evidence to suggest that the high levels of nutrients initially present in culture, e.g. glucose, and the subsequent rapid metabolism of these substrates may inhibit productivity. For example, Tovey et al. [11] showed that glucose limited steady-state chemostat cultures of mouse LS-cells grown in a chemically defined, protein-free medium produced higher titres of interferon than cultures with excess glucose. Kromer and Katinger [100] also reported that interferon production from Namalva cells growing under steady state conditions in continuous culture reversibly decreased after glucose feeding. There has been no evidence yet suggesting that excess glucose inhibits the rate of antibody production by hybridoma cells, although Low and Harbour [55] have investigated this possibility. They showed that there was no significant effect on antibody production when cells, growing in the presence of fructose, were subjected to increasing glucose concentrations.

As discussed earlier, the potential deleterious effects of excess glucose concentrations can be avoided by a variety of process control strategies including alternative carbon source, glucose pulsing in fed-batch cultures or by feeding at constant but low levels via perfusion or continuous culture systems. In contrast to these rather coarse systems, Glacken et al. [41] and Fleischaker [67] have described the application of a finely-tuned process control system. They used computer monitoring and control to optimise the production of interferon by FS-4 cells. Fleischaker [67] has described how the variations in ATP flux with time were determined by measuring oxygen uptake rate, carbon dioxide formation and lactate production. Glucose uptake was measured using an on-line spectrophotometer after automatic derivatisation of the glucose in the sample stream. The rate at which cells generated ATP was calculated by measuring the rate of glycolysis and the rate of lactate formation according to the following formula which assumes that all oxygen consumed by the cells was utilised for oxidative phosphorylation:

$$\frac{d(ATP)}{dt} = 6\,(OUR) + \frac{d(L)}{dt} \tag{9}$$

where,

OUR $=$ oxygen uptake rate
dATP/dt $=$ total rate of ATP formation
dL/dt $=$ rate of lactic acid formation

It was found that ATP flux could be used to accurately predict the cell numbers. Fleischaker [67] used the calculated ATP flux to establish a process control system specifically for glucose metabolism by FS-4 cells in order to maximise interferon production and minimise latic acid production. He observed that when glucose concentrations were below 1 mM the metabolism of FS-4 cells was similar to that found with galactose as the major carbon source, i.e. increased oxidative phosphorylation and decreased glycolysis. Thus, in order to maintain normal cell growth at a lower level of glycolysis, the glucose had to be fed at a predetermined low rate. This was achieved using three control loops, two feedback and one feedforward. The rate of lactic acid production by cells growing in the controlled glucose-fed cultures was found to be significantly lower than without control, i.e. 0.2 mmol per g dry cell per h versus 1–3 mmol per g dry cell per h. This rate was substantially reduced when the set point was lowered to 0.1 mM glucose. This excellent work has demonstrated that data gained from the measurement of the physiological processes of mammalian cells can be used to establish highly effective process control. There is no doubt that, although there is little published information currently available, this type of process will become more common as cell biologists and biochemical engineers become increasingly involved in a team approach to process optimisation.

In a recent review, Katinger [39] has described how the physiological properties of cell culture systems are influenced by the cell cultivation method, cells often behaving quite differently in closed (batch) systems compared to open (continuous) systems.

The use of a chemostat can achieve cell densities several fold higher than those obtained in batch culture. In addition, the rate of product formation is more stable in the chemostat whereas it tends to be variable during the course of a batch culture. Katinger [39] reports that his group has obtained lymphoblastoid cell densities in the order of 10^7 ml^{-1} in continuous culture, which is an extremely high yield. He also presents data which indicates that the optimal cultivation system for monoclonal antibody production is that of an immobilised cell reactor although the final concentration of the antibody and the amount of perfused media is dictated by the quality of media. This conclusion agrees with the data discussed earlier which suggested that the rate of antibody production was independent of growth rate. If, as appears from the kinetic data published to date, the antibody yield is the same at $\mu = 0$ compared to $\mu = \mu_{max}$, then product optimisation from a cell culture system exhibiting this type of production kinetics would probably be best achieved by separating the cell growth phase from the production phase. This could involve, firstly, the rapid growth of cells in a chemostat vessel and then subsequent transfer to a maintenance (production) vessel. The cells could be maintained in the second vessel in a production, non-growing mode. Appropriate maintenance levels of essential metabolites, such as glucose and oxygen, could be computed from relative data plots thus providing for an extremely economic medium supply. It is also highly likely that in this type of vessel the cells' dependence on serum could be significantly reduced, thus further reducing production costs while at the same time facilitating downstream purification procedures. This principle has apparently been successfully transferred into commercial production; i.e. it was reported that a system developed at Monsanto [101] and subsequently licensed to Invitron, was based on the finding that most cells in vivo are not normally in an active state of growth. In this process cells are first cultured in a growth vessel and then concentrated and immobilised with a finely divided, non-toxic matrix material in a static maintenance reactor. The reactor is a cylindrical vessel penetrated with an array of porous tubes for circulating medium which contains little or no serum, through the culture. The stated production yield of approximately 50 mg L^{-1} means that the 16.5 L production-scale system would be equivalent to a conventional batch system of 1,000 L. A major advantage of this type of maintenance reactor is the fact that in theory the cells can be kept in the productive non-growing mode indefinitely. For this to occur the process control would need to be extremely accurate and reliable.

5.3 Bioreactor Design

Optimal reactor design for any biochemical process involves an optimisation of a number of interrelated fundamental and practical aspects. The balance between theoretical and practical considerations is a complex and often poorly considered area of reactor design. Often the best design on purely theoretical grounds is practically either not possible or brings with it a number of attendant operating problems which lead to it not being used. Simple examples exist in the area of the growth of microbes. In ethanol production, for example, the product is generally at a level at which significant inhibition to growth or ethanol production (or both) exists. Theoretically, a plug flow reactor would be suggested if minimum fermenter volume was the sole criterion. However, at the concentration of ethanol suitable for economic production of ethanol (high concentrations favour more efficient ethanol separation) the level of carbon dioxide evolution is such that significant (if not complete) mixing is superimposed on the culture. This being the case, such plants are then generally stirred since this significant mixing occurs anyway and this greatly assists the removal of heat. Hence the reactor design most commonly used is a compromise between the theoretical and the practical.

The growth of hybridoma cells also involves such compromises and it is important to consider a large number of factors when deciding on the optimum reactor design for the growth of such cells. Product inhibition could also be a problem with murine hybridomas since feedback inhibition of antibody production has been reported [102] but this does not appear to be a general phenomenon. With respect to the growth of hybridomas, the following aspects are considered important:

1) scale of production,
2) sterility,
3) kinetics of hybridoma growth,
4) media interaction with kinetics of hybridoma growth,
5) physical constraints on hybridoma cultivation.

These are discussed briefly to illustrate their possible effects on reactor design and their role in the choice of an optimum reactor.

5.3.1 Scale of Production

Scale of production has an effect in two primary ways:

a) The scale of production of an individual cell line,
b) the variety of cell lines being commercially exploited.

With respect to the scale of production of an individual cell line, the larger the scale the more economic incentive there is to produce the antibody by continuous or perfusion culture. This is the case for not only the traditionally accepted reasons of higher productivity, ease of process control and the possibility of simpler purification processes, but also for the fact that peculiarities of the growth characteristics of a particular cell line may possibly be best suited to such a mode of operation. For example, if the antibody stability was a factor, the higher throughput of a continuous culture would mean that the time the antibody spent in a possibly sub-optimal environment would be minimised compared to batch.

Further, if for example the particular cell line of choice had a strong catabolic repression by sugars (although as yet there is no evidence for this phenomenon in hybridoma cells, as discussed in Sect. 5.2), then it may be possible to have this effect strongly reduced in continuous culture where the concentration of sugar to which the cell line is exposed is considerably less than would be the case in batch culture. The possibility of using fed-batch culture, as is the case in baker's yeast production where strong catabolite repression of respiration by sugars has been a traditionally accepted control, also exists for hybridoma cells if they are subject to similar controls. However, unlike baker's yeast, where the feed rate is controlled not for ethanol production (an inhibitory product) but for carbon dioxide and water production via respiration (non-inhibitory products) at a low but acceptable growth rate, toxic products of hybridomas (e.g. lactic acid) are likely to build up, unless unacceptably low growth rates are employed. Reactors using perfusion technique may be useful here but are associated with an attendant sterility risk.

When a number of different cell lines are being produced another factor strongly influences the reactor choice. Since the economic exploitation of cell lines producing antibodies will undoubtedly be associated with the production of a large number of different cell lines and products, there is a strong incentive for effective continuous reactors. This would greatly reduce the capital costs associated with multiple batch reactors and the scheduling problems associated with slow growth in batch culture of a number of cell lines. Particularly in a developing area, such flexibilities are of significant commercial value.

5.3.2 Sterility

Maintenance of sterility also contributes to the final reactor design chosen. While it can be shown that continuous cultures are probably the desired option, the sterility requirements of such a system, particularly considering the complexity of the growth media, are considerable. The use of growth inhibitors such as antibiotics are expensive and have unquantified effects on the kinetics of growth and antibody production of hybridoma cells. In general the simplest reactor design, with respect to seals and moving parts as well as fermenter ancilliaries, achieves the best result with respect to the maintenance of sterility. Consequently, the air-lift reactor affords a simple and effective reactor design with respect to these considerations. How this design compares to other reactors with respect to efficient oxygen transfer, mixing of nutrients and heating/cooling and what economic penalty is paid for any deficiency is yet to be fully established. Likewise, the advantages of using a perfusion system to reduce the build-up of toxic end products is balanced by a significant increase in sterility risk and the practical operating effects of this have not been discussed in detail in the literature. Generally mechanical seals have reached the level of design (through microbial reactor design) to be both effective with respect to sterile operation but also with quantitative analysis of sterility risk available.

In a reactor design where cell concentration was desirable (to be discussed), clearly accelerated internal concentration affords both an advantage with respect to cost (no large settling tanks or centrifuges, etc. being required) as well as sterility over external cell concentration.

5.3.3 Kinetics of Hybridoma Growth

With respect to the growth of hybridomas, little detailed kinetic data has been reported compared to that available for microbial systems. Two major reasons for this being the case are:

a) Obtaining complete kinetic data for a particular cell line of interest is an experimentally demanding task if carried out exhaustively. A complete kinetic analysis involves both batch and continuous experimentation. This is resource intensive and does not easily address the subject of variations between cell lines. It is then not a simple task to generalise hybridoma kinetics.
b) Commercial secrecy (necessary but problematical in obtaining a unified understanding of hybridoma kinetics). Again, commercially a compromise is necessary whereby the basic kinetic information of importance to reactor design may be gathered with the minimum of experimental work, given the reality that this may need to be undertaken for a number of different cell lines.

What may be considered as the minimum kinetic data required? Clearly, if all the basic kinetic information could be obtained from batch growth rather than continuous culture, it would be, since this method of cultivation is experimentally less demanding. By analogy with microbial cultures, batch data gives information regarding the maximum growth rate the cell is able to achieve under substrate excess. Depending on the nature of the limiting substrate, it may also give some insight into whether the desired product is a growth or non-growth associated product. However, such analyses are seldom complete. For example, if a non-energy source is a limiting substrate, then production of the product exclusively after the exhaustion of the limiting substrate would suggest non-growth associated product formation. However, disregarding some antibiotic fermentations such a delineation is not usually as clearcut. Some product formation during the growth phase gives a partly growth and partly non-growth associated metabolism during this period but only at the maximum growth rate. Similarly, product formation occurring only during growth might indicate only growth associated product formation (e.g. amino acid formation) but may equally indicate that the product is a synthesis of both precursors formed during energy formation from a limiting energy source and precursors from other media components. If the same cell were placed in a culture situation where there was a small energy formation component which was not adequate to sustain growth, then such a situation would result in product formation in the absence of growth, indicating possibly non-growth associated product formation.

In a general sense, it might be stated that very few products are totally non-growth related and most have a non-growth related product formation component as an additional (and generally predominant) contribution. For example, in baker's yeast metabolism non-growth associated sugar uptake is approximately 10–30% of the maximum sugar uptake (depending on the strain) and for lactic acid bacteria of the same order [86, 103]. Thus, most product formation is a result of growth and non-growth associated product formation. In microbial growth studies, some attempts have been made using washed suspensions to estimate the extent of non-growth associated metabolism. In this method, the culture is washed to remove all available nutrients and suspended in a buffer with only a carbon and energy source [86]. Hence the situation of non-growth associated metabolism of the carbon and energy source is simulated. This methodology has not been extended to hybridoma cells where more stringent nutrient requirements and unknown susceptibility to viability loss in the absence of nutrients may play an important role.

Furthermore most analyses of growth and non-growth associated metabolism are based on simplistic assumptions. Generally, the specific substrate uptake is divided into growth and non-growth associated components:

$$q_s = \alpha\mu + \beta \tag{10}$$

where

q_s = specific substrate uptake
[substrate] [biomass]$^{-1}$ [time]$^{-1}$ = mmol g^{-1} h^{-1}
μ = specific growth rate (h^{-1})
α = growth associated metabolic constant
[substrate] [biomass]$^{-1}$ [time]$^{-1}$ = mmol g^{-1} h^{-1}
β = non-growth associated metabolic constant
[substrate] [biomass]$^{-1}$ [time]$^{-1}$ = mmol g^{-1} h^{-1}

Onto this is superimposed a stoichiometric relationship between substrate uptake and product formation:

$$q_p = \gamma q_s \tag{11}$$

where

q_p = specific product formation
[product] [biomass]$^{-1}$ [time]$^{-1}$ = g g^{-1} h^{-1}
γ = yield constant for product formation
[product] [substrate]$^{-1}$ = g mmol^{-1}

In the analysis of batch data, only one specific growth rate is used (μ_{max}) and the yield constant for product formation is evaluated at this specific growth rate and is assumed constant. Under such conditions it may be expected that the specific product formation uptake, which may now be described by the expression

$$q_p = \alpha\gamma\mu + \beta\mu \tag{12}$$

or

$$q_p = \alpha'\mu + \beta' \tag{13}$$

can be used to evaluate the degree of growth and non-growth associated metabolism. However, such an analysis would only be valid for the maximum growth rate at a constant product yield. Whether these relationships are valid under different growth conditions, e.g. different specific growth rates, is doubtful. The specific growth rate may be changed by varying the nature (quality) of the energy substrate, the nature (quality) of the nutrients supplied or the substrate supply rate at fixed substrate quality. It is unlikely that the product spectrum or individual product yields will remain constant during such culture variations and experimentally measured product distributions at different specific growth rates (either in batch with different carbon sources, or in chemostat cultures) attest to this. In such a case a simple analysis such as expressed in Eq. (13) is generally invalid or experimentally difficult to analyse.

Consequently, it is not sound to attempt to estimate the degree of growth and non-growth associated product formation from batch studies alone, although it may provide some estimate under the condition of maximum growth rate. Continuous culture experiments are the definitive method for ascertaining the degree of growth and non-growth associated product formation at a number of growth rates.

Hence, yet again a balance or compromise is evident. To undertake a complete batch and continuous kinetic analysis for a variety of cell lines is a major task. Is there a modified sence of experiments which will approach the same goal? If an experiment with a washed suspension of hybridomas was performed (assuming this is feasible), an estimate of the non-growth associated element of the product formation may be possible. This may be compared to the maximum growth related product formation if the assumption is made that the product spectrum is not significantly variable. Given this basic information, some initial reactor design strategies might follow. If the product formation is completely or largely non-growth associated, a two-stage reactor system may be suggested whereby the first reactor is a growth reactor used to provide either sufficiently high hybridoma cell concentrations for the second reactor to be able to sustain a non-growth associated product formation, or simply hybridoma cells to which a medium deficient in a required growth nutrient is fed. This principle is being used commercially to produce bulk quantities of monoclonal antibodies in the static maintenance reactor discussed previously [101].

If the product formation is completely growth associated, then the growth rate chosen would be the maximum growth rate possible (assuming that cell yield, i.e. (biomass) $(\text{substrate})^{-1} = \text{g g}^{-1}$, and product yield, i.e. (product) $(\text{substrate})^{-1} = \text{g g}^{-1}$, remained constant and that other factors (e.g. aeration, mixing, heating/cooling were not exhibiting a dominating effect).

As previously discussed, the desired product is likely to be a product formed by both growth and non-growth associated metabolism. In such cases, the optimal operating conditions for a continuous reactor are not certain without a complete analysis of a continuous culture experiment where the following variables are measured as a minimum requirement: specific growth rate; biomass concentration; residual substrate concentration(s); product(s) concentrations and specific oxygen uptake rate. In this way the optimum operating conditions to maximise desired antibody productivity may be rationally assessed.

Can the limited number of measurements stated (i.e. batch growth data and washed suspension data) be used to give an effective reactor design? For example, if from washed suspension experiments a non-growth associated product formation rate of about 40% of the maximum rate was achieved, then by concentrating the biomass to 2.5 times the normal cell yield from the substrate a similar overall product formation may be achieved. In addition to this, since no substrate is theoretically being used for growth, the yield of product (i.e. (product) $(\text{substrate})^{-1} = \text{g g}^{-1}$) should be greater than the growing case. Hence it might be decided in such a case to operate a high density, continuous reactor where the biomass density is sufficiently high to ensure operation in the non-growth associated or maintenance region. It might also be possible to increase the maintenance (or non-growth associated) component of the substrate uptake by manipulation of the media components. This would greatly assist in reducing the cell concentration factor required to achieve the desired productivity. It should be remembered that such strategies would need to be balanced off against contamination risks and process complexities.

If the non-growth associated component is very low and cannot be increased by some method (either cell line selection or media manipulation), then the choice becomes either using the maximum growth rate and associated product formation rate from the batch data or the complete continuous culture analysis previously outlined. The compromise which must be confronted when complete continuous culture analysis is suggested is whether any possible yield or productivity increases that may be obtained from such an analysis offsets the experimental or analytical effort required. Such factors as the value of the product and its projected volume required would be important. This data would generally be available data and consequently some natural decision as to the extent of experimentation required could be made.

5.3.4 Media Interaction with Kinetics of Hybridoma Growth

In any discussion of the effect of media on hybridoma growth, again it should be appreciated that a compromise exists between the obvious advantages of a completely defined medium or at least an understanding of the major medium components influencing the growth characteristics of a hybridoma cell line on a complex (e.g. serum based) medium, and the amount of experimental work required to undertake a complete characterisation of growth requirements for a hybridoma. This compromise must further be considered with respect to the likelihood of significant differences existing between cell lines (and consequently the possible amplification of the experimental work required). In addition, a clear commercial consideration is whether any differential product yield increase affects the extensive experimentation needed to achieve it.

It is difficult to obtain from the literature an indication as to whether significant increases in antibody yield or production rate are likely to be obtained by a detailed study of the media used to grow hybridomas. While an economic incentive exists to replace serum-based media with a more defined and less expensive medium, the direct influence of this replacement on product yield is not well established. An analogy may be shown here with the aerobic growth of baker's yeast (*Saccharomyces cerevesiae*) on both a defined medium and a complex medium. It has been found that, when a defined medium is used, vigorous growth with typical cell yields can be obtained. However, depending on particular components of the defined medium, very different kinetic responses can be obtained [104, 105]. For example, in continuous culture under glucose limitation, the maximum rate of complete respiration possible and the control of the relative extents of fermentation and respiration are strongly dependent on a complete medium, particularly metallic species [104].

Two points are of importance from this work on yeast. Firstly, the recognition of the importance of these media effects only slowly developed (over a period of about 15 years). This is the result of both the extensive experimentation required to elucidate such effects and these differences being initially dismissed as strain differences or errors in experimental measurement. Hence the acceptance of overall metabolic control concepts is not an area without contention.

A similar discussion existed (still exists?) in the area of what determines the protein content of yeasts used for single-cell protein. Most studies were carried out on poorly defined, industrial media (e.g. sulphate liquors, etc.). In continuous culture studies, very many contradictions existed. For example, there was significant difference in protein content (or yield) at comparable growth rates as well as cases where protein yield increased, decreased or remained constant with increasing growth rate [106–108]. Such observations made it impossible to rationally design a reactor for single-cell protein production from a fundamental basis. It was only when work was undertaken with a carefully defined medium, that insight into the control of protein production in *Candida utilis* was achieved [109]. Whether this control strategy exists for other strains of microorganisms or for industrial media supplemented to ensure that only the carbon and energy source is limiting, has not been studied.

Hence from these two examples with yeast, it is clear that substantial experimentation is required to obtain a general understanding of the controls over the production of a desired product and the media considerations may have a profound effect on the kinetic response of the cell.

The influence of media on reactor design in hybridoma growth can then be related to the possibility of obtaining increased product yield or productivity by medium manipulation, or maintaining the product yield and productivity of the hybridoma on a complex (serum-based) medium using a less expensive defined medium. Experimental measurements have shown that changes in the medium composition may affect the product spectrum and antibody yield (see earlier Section). At present, much more experimentation is required to evaluate the effect of varying medium compositions on product yield and productivity as well as the production of toxic end products. Once the extent of experimentation has increased substantially, it then becomes necessary to put all of the experimental data into some ordered structure wherein overall control strategies can be postulated. Historically, mathematical models have been used in microbial growth to achieve this purpose. Such mathematical analysis may be as basic as the calculation of specific uptake to production rates and yields as a function of various growth conditions to a sophisticated network of the relevant metabolic processes. It is the authors' view that the large number of pathways involved in the production of energy, new cell materials and catabolic products, complex biosynthetic products and products related to both catabolism and biosynthesis, necessitates an extensive mathematical analysis. The determination of the extent of the catabolic usage of amino acids and the relative extent of the usage of the major catabolic pathways, such as the glycolytic and the pentose phosphate pathways, are simple examples where this might be of assistance.

Hence elucidating the effect of individual medium components on the kinetics of growth and product formation is helpful in obtaining an optimal reactor design. However, it comes with the attendant cost of substantial experimental work and the need to synthesise the information to obtain an understanding of overall cell control. Understanding the economic compromise will be different for commercial enterprises undertaking antibody production and laboratories undertaking fundamental studies of the kinetics of hybridoma growth.

5.3.5 Physical Constraints on Hybridoma Cultivation

Hybridoma cultivation introduces a number of unique considerations with respect to reactor design. As we have already discussed, it is commonly accepted that hybridoma cells are shear sensitive and that consequently agitation and aeration assume a dominating role. In addition, hybridoma cells generally grow very slowly so that irrespective of any shear sensitivity, poor distribution or mixing of feed and nutrients may lead to considerable viability loss. This consideration would be important both at low dilution rates in continuous culture and in high cell density reactors. This latter effect has been shown to be significant in bacterial growth studies.

In examining both of those effects separately, it is again the aim to present the theoretical considerations in addition to the practical and demonstrate the compromises that are necessarily made in a commercial reactor design.

With respect to shear sensitivity, considerable debate in the literature exists regarding the extent of shear sensitivity of hybridoma cells, the difference in such sensitivity between cell lines, the origin of the shear stress (whether from liquid, gas or liquid/gas) and whether this is of importance, and the dynamics of shear stress (i.e. time, shear stress level profiles and the effect of transient shear stress). The effect, consequently, is of considerable complexity and exhaustive experimentation is required to elucidate the mechanisms and quantify their effect.

It should be noted that neither the methodology for the measurement of shear stress effects has been completely devised nor the methodology which may be used to translate experimental measurements made on a cell line under conditions of defined shear stress into the cumulative overall effects of shear stress in the dynamic environment of any given reactor.

There are many considerations of importance with the methodology for analysing the effect of shear stress on hybridoma growth. It is not even clear which experimental apparatus should be employed since this requires the decision as to which element of a shear field has the predominant effect. Using the simplest example, when the decision that liquid shear (or shear fields generated by liquid flow) is of importance, considerable doubt could exist regarding the most appropriate experimental apparatus. A viscometer would introduce a constant liquid shear rate but there exists the possibility that hybridoma cells may align themselves into a constant shear field (due to their relative size compared to the smallest of liquid eddies) and that this experimental situation is vastly different from a mixed reactor where cells would, undoubtedly, cross shear lines and consequently be subject to transient shear stress of varying magnitude for varying time periods. Such a situation has two primary effects;

1) It introduces difficulty into suitable experimental methodology, and
2) if such difficulties can be overcome, provides a challenge in developing methodology to translate such experimental measurements into a practical input to reach design with respect to agitation and aeration strategies. (See later discussion.)

Returning to the choice of experimental apparatus, if it was decided that shear generated from gaseous input or generation was of importance, then again the experimental methodology would not be clear. Inspection of the reasons why shear originating from gas input or generation was considered controlling (and this would be a difficult enough proposition to justify given current knowledge and published experimental data), it would then be necessary to make some decision about the mechanism before an optimal experimental apparatus could be designed. For example, if gas impingement was considered controlling then an apparatus which ensured considerable and measurable impingement of the hybridoma with gas bubbles would need to be designed. If a synergistic effect between gas and liquid was postulated then some system where defined input for both origins were possible, would be desirable. Given the limitations with systems only considering liquid shear effects, this would present a considerable experimental program.

The other experimental methodology of importance is the quantification of shear stress damage. Various methods have been proposed for this measurement, viz viability, physical range and metabolic leakage. Given the previous discussion on kinetics, viability offers a theoretically attractive (simple, cost effective) method of quantification but does not allow the differentiation between growth and non-growth associated antibody production or its quantification. The main reason for this is that it is not clear whether cells which uptake the stain are catabolically non-viable, anabolically non-viable or both. The same problem exists in microbial growth (e.g. lactic acid bacteria metabolism) and the careful delineation between catabolic, anabolic or total non-viability has been demonstrated to be of both theoretical and commercial significance. Physical damage to cells is conclusive but may significantly underestimate antibody producing ability while metabolic leakage may indicate some cell damage, but a quantitative relationship between such leakage and reduction in antibody producing ability does not exist.

Clearly, a compromise then again exists between the extent of experimental work necessary to clearly elucidate the effect of shear stress and the practical necessity to commercialise antibody production.

The practical approach to the problem is to choose a reactor which will minimise the shear stress that the cell is subjected to whilst trying to maintain sufficient mixing and agitation. The basis of this choice is clearly biased towards the former consideration and the latter aspects are taken as being adequate. Recent experimental work [110)] studying mixing, suggests that this basis may well not result in optimal reactor design for air-lift reactors. As previously discussed, at relatively low feed rate, feed distribution and mixing may have a very important effect on cell viability.

Other reactor configurations, are then difficult to employ on a practical basis due to the limited theoretical and experimental basis for their use. It should be appreciated that each major reactor type is a compromise. For example, in a well mixed reactor the compromise is between good mixing and aeration weighted against possible shear damage. A perfusion system compromises between good cell/liquid contact with no problems of toxic product interaction with the hybridomas weighted against possible poor media access and cell viability loss.

To differentiate between these major reactor types, extensive experimentation is required. In a commercial operation this experimentation is balanced against the likely antibody yield increase obtainable. The fundamental study of this area is then of importance but is likely to be of a more long-term nature.

Given this situation, the other major problem when deciding on reactor design may be put in perspective. Once the extensive experimentation has been undertaken and the effect of shear stress has been quantified with respect to the interaction between the level of shear stress (however defined) and the time of exposure to this stress, a computational methodology may be employed to translate experimental measurements made on a cell line under conditions of defined shear stress into the cumulative overall effect of shear stress in the dynamic environment of any given reactor. This normally involves the mathematical technique of finite element analysis whereby liquid mixing patterns in the reactor are divided up so that the overall reactor may be divided into pockets of liquid spending defined residence times in various locations within the reactor. Upon this is superimposed the shear effects that each reactor location is subjected to. Whilst theoretically it is possible to combine this information into the overall effect of shear by analysing cumulative shear effect throughout the reactor over a given period of time, many practical problems still remain. Even for a simple mixed reactor with some gas sparging, no effective model exists for the mixing patterns within a reactor [110]. In addition, even in a reactor with no gas sparging, the effective distribution of shear forces from the stirrer mechanism is not even well established for standard stirring mechanisms in sufficient detail. The complexity of the rheological characteristic of hybridoma growth media only accentuates this aspect.

Nonetheless, ideal models can be used and are of some assistance [110] when other reactor types (e.g. air-lift) are employed. Those problems are magnified since very little fundamental experimental work has been undertaken.

6 Conclusion

During the past five years there have been significant and rapid developments in the large-scale culture of hybridoma cells. This has brought about a great increase in the amount of knowledge available concerning the cells' physical, chemical and biological requirements and this review has sought to bring together as much of the relevant information as possible. We have shown how the specific requirements of hybridoma cells have led to the development of a wide range of processes and bioreactors to culture the cells, including: stirred-tank reactors with modified stirring devices where the marine type impeller, vibromixer and rotating flexible sheets have replaced the turbine type impeller used in microbial systems; air-lift bioreactors; hollow-fibre membrane systems; and perfusion systems. Microencapsulation of the antibody secreting cells has also been developed as an alternative strategy for protecting the fragile cells from environmental stress.

It is clear from the wide range of processes that have been developed that there is no clear concensus on an appropriate optimal process development strategy and we have speculated in this review that many of the decisions concerning process development made to date have been, in the main, empirically based.

We have also attempted to examine the various problems which confront those who are attempting to optimise antibody production from hybridoma cells and have suggested that the data currently available in the literature is of very limited usefulness — the really relevant data may be known but it is currently of a largely proprietary nature. We expect this situation to change fairly rapidly as the analysis of cellular behaviour and physiology becomes more refined, leading to the development of integrated up-stream cell culture reactors with on-line computer control and full integration with downstream processing systems.

Genetic engineering technology should also play an important role in the future developments of monoclonal antibody production, with emphasis on the selection of cell strains with improved productivity in specified conditions and the possibility of generating cell lines with inserted growth factor genes thus allowing the cells to grow in more simple media. It is also likely that the use of hybridoma-type cell lines will not be restricted to the synthesis of monoclonal antibodies while, on the other hand, as a result of genetic engineering other cell systems, such as yeasts, may be used to synthesise antibody molecules for both diagnostic and therapeutic purposes.

7 References

1. Barford, J. P., Hall, R. J.: Process Biochem. *13*, 22 (1978)
2. Roels, J. A.: Energetics and Kinetics in Biotechnology, Amsterdam, Elsevier 1983
3. Stephanopoulos, G., San, K. Y.: Biotech. Bioeng. *26*, 1176 (1984)
4. Bailey, J. E., Ollis, D. F.: in Biochemical Engineering Fundamentals, Tokyo, McGraw Hill 1977
5. Monod, J. C.: Annu. Rev. Bacteriol. *3*, 371 (1949)
6. Tovey, M. G.: The cultivation of animal cells in continuous-flow culture, in: Animal Cell Biotechnology (eds. Spier, R. E., Griffiths, J. B.), 1, p. 195, London, Academic Press 1985
7. Birch, J. B., Pirt, S. J.: J. Cell Sci. *5*, 135 (1969)
8. Birch, J. B., Pirt, S. J.: ibid. *7*, 661 (1970)
9. Birch, J. B., Pirt, S. J.: ibid. *8*, 693 (1971)
10. Blaker, G. J., Pirt, S. J.: ibid. *8*, 701 (1971)
11. Tovey, M. G., Mathison, G. E., Pirt, S. J.: J. Gen. Virol. *20*, 29 (1973)
12. Tovey, M. G.: Ph. D. Thesis, University of London 1971
13. Tovey, M. G., Brouty-Boye, D., Gresser, I.: Proc. Natl. Acad. Sci. U.S.A. *72*, 2265 (1975)
14. Tovey, M. G., Brouty-Boye, D.: Exp. Cell Res. *101*, 346 (1976)
15. Tovey, M. G., Brouty-Boye, D.: J. Appl. Chem. Biotechnol. *26*, 345 (1976)
16. Tovey, M. G., Brouty-Boye, D.: Exp. Cell Res. *118*, 383 (1979)
17. Tovey, M. G.: Adv. Cancer Res. *33*, 1 (1980)
18. Tovey, M. G.: Use of the chemostat culture for study of the effect of interferon on tumour cell multiplication, in: Methods in Enzymology (ed. Pestka, S.), *79*, p. 391, New York, Academic Press 1981
19. Herbert, D., Elsworth, R. E., Telling, R. C.: J. Gen. Microbiol. *14*, 601 (1956)
20. Hu, W-S., Wang, D. I. C.: Mammalian cell culture technology: a review from an engineering perspective, in: Mammalian Cell Technology (ed. Thilly, W. G.), p. 167, Boston, Butterworths 1986
21. Fazekas de St. Groth, S. J.: J. Immunol. Methods, *57*, 121 (1983)
22. Boraston, R., Thomson, P. W., Garland, S., Birch, J. B.: Develop. biol. Standard., *55*, 103 (1984)
23. Velez, D., Reuveny, S., Miller, L., Macmillan, J. B.: J. Immunol. Methods, *86*, 45 (1986)
24. Lavery, M., Kearns, M. J., Price, D. G., Emery, A. N., Jeffries, R., Nienow, A. W.: Devel. biol. Standard *60*, 199 (1985)
25. Low, K.: Ph. D. Thesis, University of Sydney 1987
26. Birch, J. R., Thompson, P. W., Lambert, K., Boraston, R.: Proc. Am. Chemical Soc., Philadelphia 1984
27. Käppeli, O., Sonnleitner, B.: Critical reviews in biotech. *4*, 299 (1986)
28. Telling, R. C., Radlett, P. J.: Adv. Appl. Microbiol. *13*, 91 (1970)
29. Hirtenstein, M. D., Clark, J.: Tissue culture in medical research (eds. Richards, R., Rajan, K.), Oxford, Pergamon Press 1980
30. Cherry, R. S., Papoutsakis, E. T.: Bioprocess Eng. *1*, 29 (1986)
31. Smith, C. G., Greenfield, P. F., Randerson, D. H.: Proc. VII Australian Biotech. Conf., p 269 Melbourne 1986

32. Midler, M. Jr., Finn, R. K.: Biotech. Bioeng. *8*, 71 (1966)
33. Augenstein, D. C., Sinskey, A. J., Wang, D. I. C.: ibid. *13*, 409 (1971)
34. Leverett, I. B., Hellums, J. D., Alfrey, C. P., Lynch, E. C.: Biophysical J. *12*, 257 (1972)
35. Stathopoulos, N. A., Hellums, J. B.: Biotech. Bioeng. *27*, 1021 (1985)
36. Handa, A., Emery, A. N., Spier, R. E.: Devel. biol. Standard (in press)
37. Fleischaker, R. J., Jr., Sinskey, A.: Eur. J. Appl. Microbiol. Biotechnol. *12*, 193 (1981)
38. Spier, R. E., Griffiths, J. B.: Devel biol. Standard *55*, 81 (1984)
39. Katinger, H.: ibid. (in press)
40. Fleischaker, R. J.: Ph. D. Thesis, M. I. T. 1982
41. Glacken, M. W., Fleischaker, R. J., Sinskey, A. J.: Trends in Biotechnol. *1*, 102 (1983)
42. Acton, P. T., Lynn, J. D.: Adv. Biochem. Eng. *7*, 85 (1977)
43. Brosemer, R. W., Rutter, W. J.: Expl. Cell Res. *25*, 101 (1961)
44. Kilburn, D. G., Lilly, M. D., Webb, F. C.: J. Cell Sci. *4*, 645 (1969)
45. Mizrahi, A., Vosoeller, G. V., Yagi, Y., Moore, G. E.: P.S.E.B.M *139*, 118 (1972)
46. Reuveny, S., Velez, D., Macmillan, J. D., Miller, L.: J. Immunol. Methods, *86*, 53 (1986)
47. Reuveny, S., Velez, D., Miller, L. and Macmillan, J. D.: ibid. *86*, 61 (1986)
48. Mateles, R. I., Battat, E.: Appli. Microbiol. *28*, 901 (1974)
49. Goldberg, I., Er-el, Z.: Process Biochemistry *16*, 2 (1981)
50. Fiechter, A.: Physical and chemical parameters of microbial growth, in: Advances in Biochemical Engineering/Biotechnology (ed. Fiechter, A.) *30*, p 7, Berlin, Springer-Verlag 1984
51. Cail, R. G., Barford, J. P., Lichacz, R.: Agri. Wastes *18*, 27 (1986)
52. Reitzer, L. J., Wicc, B. M., Kennell, D.: J. Biol. Chem. *254*, 2669 (1979)
53. Zielke, H. R., Ozand, P. T., Tildon, J. T., Sevdalian, D. A., Cornblath, M.: J. Cell Physiol, *95*, 41 (1978)
54. Low, K., Harbour, C., Fletcher, A.: Proc. VI Australian Biotech. Conf. Brisbane, 1984
55. Low, K., Harbour, C.: Develop. biol. Standard. *60*, 73 (1985b)
56. Thomas, J. N.: Nutrients, Oxygen and pH, in: Mammalian Cell Technology (ed. Thilly, W. G.), p 109, Boston, Butterworths 1986
57. Lambert. K., Pirt, S. J.: J. Cell Sci. *17*, 397 (1975)
58. Arathon, W. R., Telling, R. C.: Develop. biol. Standard. *50*, 145 (1982)
59. Butler, M., Imamura, T., Thomas, J., Thilly, W. G.: J. Cell Sci. *61*, 351 (1983)
60. Butler, M.: Develop. biol. Standard. *60*, 269, (1985)
61. Roberts, R. S., Hsu H. W., Lin, K. D., Yang, T.: J. Cell Sci. *21*, 609 (1976)
62. Stoner, G. D., Merchant, D. J.: In vitro *7*, 330 (1972)
63. Griffiths, J. B., Pirt, S. J.: Proc. Roy. Soc. *B168*, 421 (1967)
64. Holley, R. W., Armour, R., Baldwin, J. H.: Proc. Nat. Acad. Sci. U.S.A. *75*, 1864 (1978)
65. Visk, W. J., Kolodny, G. M., Gross, P. R.: J. Cell Physiol. *80*, 373 (1972)
66. Eagle, H., Barban, S., Levy, M., Schulze, H. O.: J. Biol. Chem. *233*, 551 (1958)
67. Fleischaker, R. J. Jr.: Practical matters in instrumentation for mammalian cell cultures, in: Mammalian Cell Technology (ed. Thilly, W. G.), p. 199, Boston, Butterworths 1986
68. Cristofalo, V. J., Kritchevsky, D.: J. Cell Physiol. *67*, 125 (1966)
69. Imamura, T., Crespi, C. L., Thilly, W. G., Brunengraber, H.: Anal. Biochem. *124*, 353 (1982)
70. Kilburn, D. G., and Webb, F. C.: Biotech. Bioeng. *10*, 801 (1968)
71. Mizrahi, A.: Develop. biol. Standard. *55*, 93 (1984)
72. Higuchi, K.: Adv. Appli. Microbiol. *16*, 111 (1973)
73. Ham, R. G., McKeehan, W. L.: Media and growth requirements, in: Methods in Enzymology, *58*, (eds. Jakoby, W. B., Pastan, I. H.), p. 44, New York, Academic Press 1979
74. Lambert, K. J., Birch, J. R.: Cell growth media, in: Animal Cell Biotechnology (eds. Spier, R. E., Griffiths, J. B.), *1*, p. 85 London, Academic Press 1985
75. Barnes, D., Sato, G.: Cell *22*, 649 (1980)
76. Chang, T. H., Steplewski, Z., Koprowski, H.: J. Immunol. Methods, *39*, 369 (1980)
77. Darfler, F. J., Insel, P. A.: J. Cell Physiol. *115*, 31 (1983)
78. Murakami, H., Masui, H., Sato, G. H., Sneko, N], Chow, T. P., Kano-Sueoka, T.: Proc. Nat. Acad. Sci. U.S.A. *79*, 1158 (1982)
79. Murakami, H., Masui, H., Sato, G. H.: Suspension culture of hybridoma cells in: serum-free medium: soybean phospholipids as the essential components, in: Growth of Cells in hormonally

defined media (eds. Sato, G. H., Pardee, A. B., Sirbasku, D. A.) p. 711, New York, Cold Spring Harbour 1982
80. Kawamoto, T., Sato, J. D., Le, A., McClure, D. B., Sato, G. H.: Anal. Biochem. *130*, 445 (1983)
81. Iscove, N. N., Melchers, F.: J. Exp. Med. *147*, 923, (1978)
82. Low, K., Harbour, C.: Develop. biol. Standard. *60*, 17 (1985a)
83. McAteer, J. A., Douglas, W. H. J.: Monolayer culture techniques in: Methods in Enzymology (eds. Jakoky, W. B., Paston, I. H.), *58*, p. 1315 London, Academic Press 1979
84. Himmler, G., Palfi, G., Riiker, F., Katinger, H., Scheirer, W.: Develop, biol. Standard. *60*, 291 (1985)
85. Shaw, M. K.: J. Bacteriol. *93*, 1332 (1967)
86. Franks, P. A., Hall, R. J., Linklater, P. M.: Biotech. Bioeng. *22*, 1465 (1980)
87. Rubin, H.: J. Cell Biol. *51*, 686 (1971)
88. Eagle, H.: J. Cell Physiol. *82*, 1 (1973)
89. Barton, M. E.: Biotech. Bioeng. *13*, 471 (1971)
90. Ceccarini, C., Eagle, M.: Proc. Nat. Acad. Sci. U.S.A. *68*, 229 (1971)
91. Birch, J. R., Edwards, D. J.: Develop. biol. Standard. *46*, 59 (1980)
92. Taylor, W. G., Lamalier, R. F., Sanford, K. K.: J. Cell Physiol. *95*, 33 (1978)
93. Kilburn, D. G., Lilly, M. D., Self, D. A., Webb, F. C.: J. Cell Sci. *4*, 25 (1969)
94. Kozbor, D., Lagarde, A. E., Roder, J. C.: Proc. Natl. Acad. Sci. U.S.A. *79*, 6651 (1982)
95. Bron, D., Feinberg, M. B., Teng, N. N. H., Kaplan, H. S.: ibid. *81*, 3214 (1984)
96. Foung, S. K. H., Perkins, S., Raubitschek, A., Larrick, J., Lizak, G., Fishwild, D., Engelman, E. G., Grumet, F. C.: J. Immunol. Methods *70*, 83 (1984)
97. Harris, J. L., Spier, R. E.: Physical and chemical parameters: measurement and control, in: Animal Cell Biotechnology (eds. Spier, R. E., Griffiths, J. B.) *1*, p. 283, London, Academic Press 1985
98. Harris, J. L., Spier, R. E.: Computer applications in animal cell biotechnology, in: Animal Cell Biotechnology (eds. Spier, R. E., Griffiths, J. B.) *1*, p. 321 (1985)
99. Van Wezel, A. L., van der Velden-de Groot, C. A. M., de Haan, H. H., van den Heuvel, N., Schasfoort, R.: Devel. biol. Stand. *60*, 229 (1985)
100. Kromer, E., Katinger, H. W. D.: ibid. *50*, 349 (1982)
101. Van Brunt, J.: Biotechnology *4*, 505 (1986)
102. Merten, O. W., Reiter, S., Himmler, G., Scheirer, W., Katinger, K.: Develop. biol. Standard. *60*, 219 (1985)
103. Barford, J. P., Hall, R. J.: Biotech. Bioeng. *23*, 1735 (1981)
104. Rieger, M., Käppeli, O., Fiechter, A.: J. Gen Microbiol. *129*, 653 (1983)
105. Barford, J. P., Jeffrey, P. M., Hall, R. J.: The Crabtree effect in *Saccharomyces cerevisiae* — primary control or transient?, in: Advances in Biotechnology, *1*, p. 255 (eds. Moo-Young, M., Robinson, C. W., Vezina, C.) London, Pergamon Press 1980
106. Herbert, D.: Symp. Soc. Gen. Microbiol. *11*, 391 (1961)
107. Abbott, B. J., Laskin, A. I., McCoy, C. J.: Appl. Microbiol. *28*, 58 (1974)
108. Cooney, C. L., Wang, D. I. C., Mateles, R. I.: Appl. Environ. Microbiol. *31*, 91 (1976)
109. Schwartzkoff, C. L., Rogers, P. L.: J. Gen. Microbiol. *128*, 1635 (1982)
110. Mann, R.: Gas-Liquid Contacting in Mixing Vessels, Rugby, Institution of Chemical Engineers 1983

The Use of Plasma Protein Fractions as Medium Supplements for Animal Cell Culture

A. J. MacLeod
Scottish National Blood Transfusion Service Protein Fractionation Centre, Edinburgh, EH17 7QT, G.B.

Plasma and serum protein fractions have been investigated as potential sources of protein supplement for animal cell culture media by several workers since the 1950s. The results however have been often confused and sometimes conflicting. Recently using very careful control of the processing conditions, especially temperature and pH, it has been found that it is possible to use Cohn fraction IV as a source of medium supplement although it appears that there is some variation in the properties of Cohn fractions prepared by different manufacturers. This development opens the way to cell culture for the production of proteins for clinical use using a nutrient medium containing proteins exclusively of human origin. While reducing the demands made on down-stream purification the use of these fractions introduces new problems of possible infectious contamination which will make it essential that all cell products produced using these supplements are submitted to an effective virus inactivation procedure.

1 Introduction

Recent developments in the manipulation of the genetic constitution of animal cells and in their maintenance in large-scale culture have opened a way to production on demand of native proteins in particular for clinical use with human patients. However unlike the use of vaccines, with which parallels often are drawn, the recipients of many of the proposed products will not be healthy, as in cases of thrombosis or cirrhosis, or they receive repeated treatment throughout their lives, as in cases of haemophilia. Thus they may be particularly exposed and vulnerable to an adverse reaction to even trace contaminants in a product. This is especially the case if the contamination is heterologous protein as experience with human plasma protein preparations already well established in routine clinical practice for treatment of some of these conditions suggests that redundant human protein in the product can be tolerated by patients much better than comparable quantities of non-human protein.

Nutrient media for maintenance of animal cells in vitro are usually aqueous solutions of a number of chemically defined low molecular weight components such as amino acids, carbohydrate, vitamins and salts, to which normally it is necessary to add a protein supplement such as serum.

In the production of proteins using animal cell culture systems the sources of redundant protein in the final product will be the cells and the protein supplement added to the medium. Consequently when the product is intended for clinical use it would be preferable if both cells and supplement could be of human origin. An immediate advantage of being able to use such an approach would be a dramatic reduction in the complexity and cost of down-stream processing because it would no longer be necessary to try to achieve an absolutely pure product. However, human serum is very scarce and expensive and would not be available in quantities sufficient to sustain an industrial cell culture process. Human protein preparations are produced on an industrial scale using plasma supplied by donors and one consequence of this established fractionation process is that some by-products are generated which are of little clinical value but which can be used to produce a proteinaceous culture medium supplement suitable for use with some important cell types [1)].

The serum supplement of cell culture medium has a number of functions which can be broadly categorised [2)]:

— Attachment and spreading.
— Nutrition (trace elements, labile or water insoluble nutrients).
— Stimulation (growth factors, hormones, proteases).
— Protection: a) biological (antitoxin, antioxidant, antiprotease).
b) mechanical in agitated systems.

It has been clear for a long time that it is necessary to optimize the formulation of the defined components of media to be used for different types of cells, but only relatively recently has substantial progress been made in isolating the constituents of serum, identifying their function and optimising their formulation for in vitro cultivation of different types of cells.

A requirement for a particular protein in the medium varies with the type of cell to be cultured. Some cell lines such as Namalva can be grown satisfactorily in medium in which the only protein is albumin [3)]. Many other cell lines have been found to require several supplemental proteins, typically including albumin, transferrin and insulin with the possible addition of polypeptide growth factors which may be isolated from non-serum sources and which have been shown to stimulate a variety of cell types in in vitro culture [4)].

Many completely defined media are available now which contain only highly purified proteins and which may be used for large-scale animal cell culture. However although the protein content of serum-free media is usually very much lower than that of serum supplemented media, which makes down-stream processing to recover a pure product much easier, they are significantly more expensive, many still contain non-human protein and in general it must be remarked that any move away from serum as the medium supplement will tend to emphasise differences between the requirements of different types of cells [5,6]. Thus there is apparently a choice to be made between serum, a versatile, relatively cheap supplement, and serum-free media which are more specific in their application and in most cases are much more expensive but offer considerable savings in the complexity and cost of down-stream processing and which are not as liable to fluctuations in quality and availability.

Human plasma protein fractions represent an alternative for the production of proteins for clinical use with some of the advantages of both serum-based and serum-free medium options. The fractions are complex mixtures of proteins and show some of the multifunctional versatility of serum coupled with the prospect of considerably reduced complexity and cost of down-stream processing. The ability to use by-products of routine human plasma fractionation to sustain animal cell cultures would also increase the output of the existing fractionation industry thereby improving its efficiency and contributing further to the overall economy of the two operations.

2 Properties of Plasma and Serum

The medium supplement used in conventional in vitro animal cell culture is serum obtained from whole blood. Whole blood consists of formed elements, red cells, white cells and platelets, suspended in a proteinaceous fluid, the plasma. In an adult human the plasma normally accounts for about 80% of the blood volume and contains about 70 g of total protein per litre. Plasma contains hundreds of different proteins, the major components are listed in Table 1 [7]. Plasma can be collected by addition of an anticoagulant such as citrate, heparin or EDTA to the whole blood which is then centrifuged to pack the cells so that most of the plasma can be decanted. The extent to which the plasma is free of platelets in particular depends on the conditions used for this separation.

Table 1. Some major plasma proteins

Protein	Approximate quantity in normal human plasma g l^{-1}
Albumin	35–55
Immunoglobulin G	8–18
Immunoglobulin A	0.8–4.5
Fibrinogen	3
Transferrin	3
α_1-Antiprotease	3
α_2-Macroglobulin	2.5
Immunoglobulin M	0.6–2.5
Haptoglobin	1.7–2.3
Fibronectin	0.3

Serum is the liquid that can be collected by allowing whole blood to clot and removing the formed elements along with the coagulated protein. Serum has a very similar protein composition to plasma from which it is derived. There are a number of specific differences between plasma and serum however, of which the most obvious is the presence in plasma of fibrinogen. The production of the clot in whole blood is a result of a complex interaction of many components of the plasma which eventually result in the conversion of circulating fibrinogen to fibrin by the action of the serine protease thrombin. The fibrin then polymerises to form a complex structure, the fibrin clot. The coagulation process however involves many steps in a cascade reaction prior to the activation of the protease zymogen prothrombin. There are also other consequences of activation of the coagulation process which in vivo make the initial wound closure more secure by further cross-linking the molecules of the fibrin clot, restrict the coagulation process to the vicinity of the wound, attack organisms that infect the wound, promote cell growth to heal the wound and eventually digest the clot to clean up the site of the wound [8]. Consequently it is not correct to regard serum simply as defibrinated plasma. Many molecules are activated or released during serum production some of which have profound effects on biological systems in trace amounts and these include a number of cell growth factors.

2.1 Platelet-derived Growth Factor

An important growth factor found in serum is the platelet-derived growth factor (PDGF). This is stored in the α-granules of platelets and is released after platelet activation for instance by the action of thrombin [9]. Thus PDGF is released into serum produced from whole blood but it is only present in trace amounts in plasma and serum produced from platelet-poor plasma. The discovery that part of the amino acid sequence of PDGF is very similar to that of the transforming protein of the Simian Sarcoma Virus [10] and a number of oncogene products [11] indicates the general nature of the mitogenic stimulus that may be provided by this protein. However there has been some controversy over the role of PDGF in stimulating animal cells in vitro [12] and it is possible to grow normal cells apparently in the absence of PDGF [13].

2.2 Insulin-like Growth Factors

The Insulin-like Growth Factors (ILGF) constitute a second group of growth factors and these can be detected in both serum and plasma. ILGF are members of a group of polypeptide hormones that are characterised by their structural relationship to insulin [14]. Typically they have a molecular weight of about 8000 and circulate in the plasma bound to large carrier-proteins [15]. One consequence of this arrangement is that ILGF have much longer half-lives in the circulation than polypeptide hormones which exist in the plasma in an unassociated form. ILGF can be released from their carrier protein either by limited proteolytic activity or by extraction at acid pH. In plasma ILGF can be detected in complexes with a molecular weight of about 50,000 or about 150,000 but in serum only the larger of these is found. This leads to the proposal that the carrier- protein/ILGF complex exists in an equilibrium between single and triple forms but that the ILGF can only be released from the single form. ILGF release as a consequence of proteolytic digestion of its complex with the carrier-protein will occur during the coagulation process in vivo or serum production in vitro and the 50,000 molecular weight single form of the complex will be depleted. Normal human skin fibroblasts can be grown in medium supplemented with platelet-poor plasma but this is only possible if the calcium ion level is about 1.8 mM. If the level of calcium ion is reduced to 0.2 mM these cells will no longer grow, suggesting the possibility that a calcium ion-dependent protease releases the ILGF [16].

2.3 Growth Factor Interactions

In vivo all the factors generated by activation of the response to a wound, including coagulation, will interact and clear syngergies between ILGF and PDGF [17] and between ILGF and steroid hormones [18] have been demonstrated. Thus although any one of these factors may be able to induce cell growth and division in vitro the response to them individually by normal cells is unlikely to be optimal.

2.4 Fibrinogen and Fibronectin

Other proteins which apparently have major roles in sustaining animal cells in vitro such as albumin and transferrin seem to be present in the same form in both plasma and serum as do other components such as lipids, trace elements and reducing agents. However to prevent formation of a clot it is necessary to defibrinate plasma before it can be used to supplement cell culture medium. A number of methods of defibrination can be used including production of serum by recalcification of the plasma [19]. Other methods such as heating plasma at 57 °C for 35 min have been used successfully [12] but some methods such as cryoprecipitation or adsorption onto inorganic compounds such as kaolin yield a product which will support cells that grow in suspension [20] but not those which need to attach to a surface. This is probably due to removal of fibronectin along with the fibrinogen by these methods. Fibronectin has been shown to be relatively stable at 57 °C and would not be affected by heat defibrination [21].

2.5 Summary

Whole blood is a very convenient source of complex factors necessary for in vitro cultivation of animal cells. However the blood must be processed to make these factors available in a form which it is practicable to use. The very first step in this processing, recovery of plasma or serum, will have profound effects on some extremely potent components which will alter their behaviour in subsequent fractionation steps. Consequently it is important in all discussions of the use of plasma fractions as cell culture medium supplements to consider whether the fractions are truly of plasma or are in fact of serum proteins.

3 Plasma Protein Fractionation

The origins of human blood plasma fractionation to provide protein concentrates for clinical use, lie in the work of E. J. Cohn and his co-workers at the Harvard University Medical School in the 1940's. The initial objective of Cohn's group was the preparation of a blood volume replacement solution that could be used under military conditions. The resultant product was a solution of human plasma proteins, substantially albumin, and the method of preparation involved fractional precipitation with cold ethanol now known as Cohn Fractionation [22].

Five variables were identified by Cohn's group which could be used to separate plasma proteins according to differences in their solubility behaviour. These variables were:

- dielectric constant,
- pH,
- temperature,
- ionic strength,
- protein concentration.

Given the nature of the plasma feedstock there is a limit to the practicability of modifying ionic strength or protein concentration and consequently processes that have been developed using this chemistry depend primarily on altering the dielectric constant, pH and temperature. Dielectric constant is changed by addition of an organic solvent, now almost universally ethanol, and pH by addition of acid [23]. The main features of the process can be described by plotting ethanol concentration against pH and in practise these two are changed in a number of steps to give a series of discrete protein fractions (Fig. 1). However ethanol will readily denature proteins and has an appreciable positive heat of solution in aqueous systems. Consequently the process

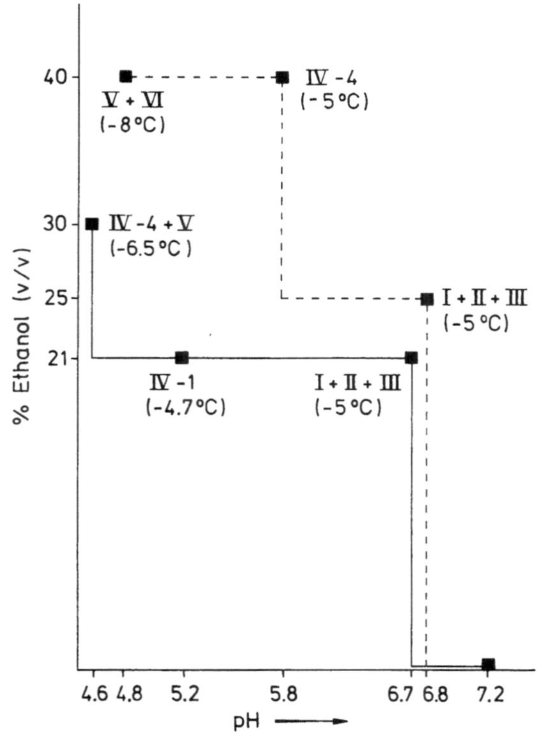

Fig. 1. Fractionation of human plasma by the cold ethanol process, as carried out at the Protein Fractionation Centre, Scottish National Blood Transfusion Service

Table 2. Distribution of some major plasma proteins in Cohn cold ethanol fractionation of normal human plasma

Fractions	I	II	III	IV-1	IV-4	V
Major proteins present	Fibrinogen fibronectin	Immunoglobulin	α_2-Macroglobulin	α_1-Antiprotease antithrombin III transferrin caeruloplasmin	α_1-Antiprotease transferrin caeruloplasmin haptoglobin	Albumin
Approximate proportion (% by wt of total protein)	5.5		25	6	7	55

Proteins, notably albumin, from occluded liquor are found in each of these fractions is normally carried out at temperatures of about —5 °C and below. It is vital that the addition of ethanol to plasma is very carefully controlled with thorough mixing both to prevent local high concentrations of ethanol occuring and to remove the heat generated. It was with the objective of improving the mixing and control of the process that the Continuous Small Volume Mixing system used at the Protein Fractionation Centre of the Scottish National Blood Transfusion Service was developed [24].

Cohn published details of several different versions of his plasma fractionation process, most of the procedures now in large-scale use being derived from his method

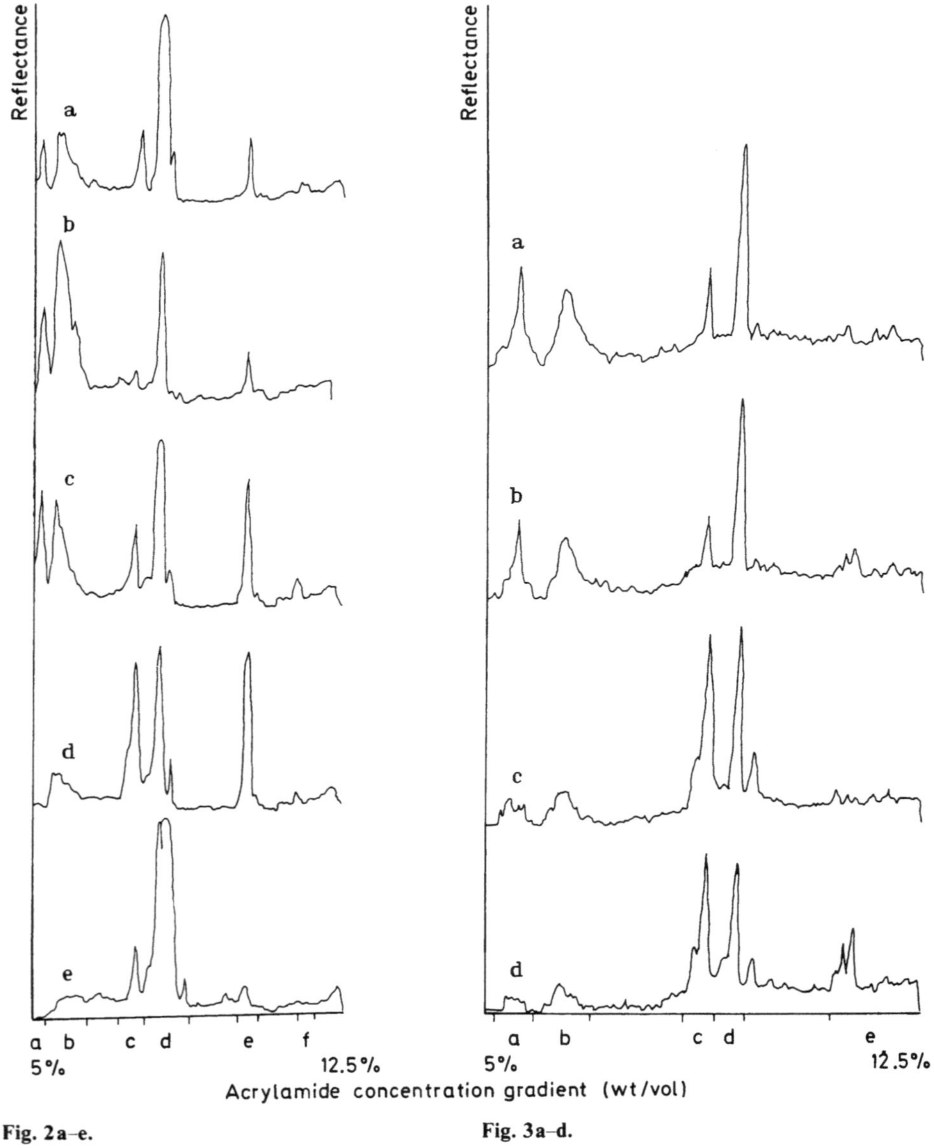

Fig. 2a–e. **Fig. 3a–d.**

six. Five major fractions were identified that could be collected using cold-ethanol precipitation and they can be characterised by their major constituents (Table 2) [7] and can be distinguished by analysis using, for instance, polyacrylamide gel electrophoresis (Fig. 2). There is a small amount of protein left in the supernatant following removal of fraction V but this is less than 2% of the protein in the plasma feedstock and subsequent fractions are very small.

Within the conditions described by Cohn for precipitation of plasma protein fractions there is an infinite number of possible combinations of ethanol concentration and pH. Thus a particular type of fraction may be precipitated by several different combinations of conditions. This has been exploited by the major manufacturers of blood products to tailor the fractionation process to suit their individual requirements. In the case of fraction IV especially a wide range of conditions will precipitate broadly similar fractions and consequently sub-fractions have been identified within fraction IV, those collected at the Protein Fractionation Centre are designated IV-1 and IV-4. Which version of fraction IV is collected depends on the specific objective of the process. The most obvious difference between fractions IV-1 and IV-4 is that the latter has a substantially higher content of albumin and transferrin and a much lower content of γ-globulin and very high molecular weight material as can be seen in Fig. 2. Until recently these fractions have been discarded as a by-product of the fractionation process and the requirements for yield or purity of other fractions have dictated the conditions for their collection.

4 Use of Plasma Fractions in Cell Culture

Considerable variation has been found in the response of cells to different batches of serum. There have been several investigations into this batch-to-batch variability and the Cohn cold-ethanol fractionation process has been used in some attempts to at

◀ **Fig. 2a–e.** Analysis of normal human plasma and of Cohn fractions by polyacrylamide gel electrophoresis in a 5% to 12.5% (w/v) acrylamide concentration gradient using the method of Laemmli [25]. The gels were stained with Coomassie Brillant Blue, dried and scanned in a reflectance densitometer; **a** normal human plasma; **b** Cohn fraction I + II + III; **c** Cohn fraction IV-1; **d** Cohn fraction IV-4; **e** Cohn fraction IV-4 + V. The major components of the peaks indicated are: (a) very large molecular weight material >250,000 Da; (b) immunoglobulin G; (c) transferrin; (d) albumin; (e) low molecular weight material 15,000–20,000 Da; (f) very low molecular weight material 5000–10,000 Da

Fig. 3a–d. Cohn fractions IV-1 and IV-4 processed either by the method of Ng and Dobkin [41] or by that of MacLeod [42]. The samples were analysed by polycrylamide gel electrophoresis in a 5% to 12.5% (w/v) acrylamide concentration gradient using the method of Laemmli [25]. The gels were stained with Coomassie Brilliant Blue, dried and scanned in a reflectance densitometer: **a** fraction IV-1, method of Ng and Dobkin; **b** fraction IV-1, method of MacLeod; **c** fraction IV-4, method of Ng and Dobkin; **d** fraction IV-4, method of MacLeod. The major components of the peaks indicated are: (a) very large molecular weight material >250,000 Da; (b) immunoglobulin G; (c) transferrin; (d) albumin; (e) low molecular weight material, 5000–20,000 Da least partially purify and characterise factors that stimulate or inhibit cell growth. Reports of work in this area date back to the mid-1950's [26, 27].

A major problem in assessing this literature is that critical information is frequently missing or is presented ambiguously. There is particular confusion over the use of plasma or serum. As explained above the biochemical differences between plasma and serum extend far beyond the simple presence or absence of fibrinogen. An example of this problem is to be found in the paper of Chang et al. [28] where information is given for preparation and fractionation of both plasma and serum but the results refer only to serum. Similarly the paper of De Luca et al. [29] specifically uses the word "plasma" in the title but refers exclusively to serum in the text.

Typically about 30% to 40% of the mass of the collected fraction is precipitated protein, the remainder being occluded liquor. In the case of fractions collected at the Protein Fractionation Centre (Fig. 1) this liquor is 21% (v/v) ethanol and pH 5.2 in the case of IV-1 or 40% (v/v) ethanol and pH 5.85 in the case of IV-4. If these pastes are allowed to warm to above +4 °C extensive irreversible denaturation occurs. They also have a high salt content and substantial buffering capacity. A consequence of this is evident in the paper of De Luca et al. [30] where extraction with saline is described but it is noted that only about 30% of the protein dissolved. This paper is unusual in that it gives even this amount of detail of the extraction method. In most instances the collection, handling and extraction of the Cohn fraction is only vaguely described.

Confusion over biochemical details is compounded by the variety of cell systems that have been used to evaluate extracts of Cohn fractions. Many authors have reported detection of cell growth inhibitory activity in Cohn fraction IV-1 but there is little agreement as to the cause. Chang et al. [28] merely recorded that Hela cells and normal human conjunctival cells degenerated in medium supplemented with fraction IV-1 material in contrast to their vigorous growth when fraction IV-4 was used. De Luca et al. [30] working with RPMI 2402 hamster carcinoma cells identified an inhibitor activity in bovine fraction IV-1 apparently produced from serum, but this was not detected in all preparations and not in any derived from human serum. Complement was ruled out and later detergent activity of free fatty acids and a polypeptide factor were proposed as the cause of the growth inhibition [29].

Pirt and Lambert [31] noted that some cell growth factor activity for normal human diploid cells could be detected in Cohn fraction IV, but they found the method poorly reproducible and some serum was required in the culture medium. The poor reproducibility of the method may reflect the difficulty of controlling the cold ethanol precipitation process, but as no details of the methods are given it is not possible to identify the actual nature of the product.

Melnick and Wallis [32] using primary cultures of monkey kidney cells identified antiprotease activity as an important component in the role of serum in cell culture medium. They found that Cohn fraction IV-4 or fetuin could replace serum in their culture system whereas Cohn fraction IV or IV-1 could not and that this correlated with the protease inhibiting activity of each of the preparations. Again no details of the preparation of the Cohn fractions were provided. Recently α2-macroglobulin, a potent protease inhibitor, has been identified as the active component of fetuin [33].

Spieker-Polet et al. [34] studied the response of mouse and rat lymphocytes to stimulation with the mitogen Concanavalin A in medium supplemented with each of the Cohn fractions. Little or no response was seen in protein-free medium or when the medium was supplemented with fractions II, III or IV. Supplementation of the basal medium with fraction VI gave a significant response but by far the greatest effect was obtained when albumin, fraction V, was used. This was attributed, primarily, to the nutrient value of free fatty acids bound to the albumin.

A requirement for completely defined cell culture systems coupled with variable quality and availability of satisfactory animal serum stimulated a great deal of interest in the identification of the components of serum necessary to support animal cell growth and function in vitro. Iscove and Melchers [35] published a formulation developed to support peripheral blood lymphocytes in vitro, in which the only proteins were albumin and transferrin. Insulin had earlier been identified as a growth factor for some types of cells [36] and these three proteins now provide a common basis to many of the serum free formulations that have been published [4]. Albumin and transferrin are major components of all forms of Cohn fraction IV and in addition fraction IV-1 has been recognised as a ready source of the Insulin-like Growth Factors [37]. This combination of components suggests that Cohn fraction IV-1 could provide a possible source of a substitute for whole serum in culture medium on the one hand or of highly purified, and expensive, proteins on the other [38]. However recent papers mentioning the possible use of Cohn fraction IV have tended to regard it exclusively as a source of Insulin-like Growth Factors [39, 40].

4.1 Fraction IV and Sub-fractions

Recently details have been published of two contrasting approaches to the exploitation of Cohn fractions as protein supplements for cell culture medium, developed by Ng and Dobkin [41] and by MacLeod [42].

The former method involves resuspending the protein paste in water, centrifuging to remove insoluble material, adjusting the pH of the supernatant to 6.9, dialysing against 2000 volumes of 0.85% saline and sterilising by 0.22 µm filtration. A critical feature of this process is the actual volume of dialysis fluid used which if essential would itself make the product twice as expensive per unit volume as foetal calf serum. The yield is about 120 g of protein per kg of fraction IV paste. The product has a total protein content of 30 g l^{-1} and is used at 5% (v/v) of the final culture medium.

It is claimed that cell growth inhibiting substances, in the molecular weight range of about 2.5×10^5 to 1.0×10^{10}, are left in the solid material which is removed by centrifugation. Cell growth inhibitors of this size have been identified in serum and shown to be an α_2-lipoprotein [43] which would be expected to separate in Cohn fraction IV [44], or high molecular weight α_2-globulin which is not a lipoprotein [45] but which may also separate in Cohn fraction IV. Thus it is conceivable that a process to remove high molecular weight material selectively could indeed enhance the suitability of Cohn fraction IV as a cell culture medium supplement. However in this case it is curious that although fraction IV-4 yields a product that supports cell growth it was found that fraction IV-1 did not. The cells used to evaluate these products were the lymphoblastoid Namalva line. It has been established that Namalva cells will grow in medium supplemented only with albumin [3] and the data provided by Ng and Dobkin demonstrates the same phenomenon although growth is slightly increased when the fraction IV extract is used.

The process of MacLeod differs from that of Ng and Dobkin in that the objective is to redissolve as much of the protein as possible from the Cohn fraction paste. This is achieved by adding a little water to the paste to make a free flowing slurry the pH of which is raised to 7.2, the appearance of the fraction IV-1 slurry in particular changing dramatically at about pH 6.5 as the proteins redissolve. The solution is diluted to give a total protein content of 12 g l^{-1} and is then centrifuged. Fraction IV-1 produces a very small firm translucent pellet and the supernatant is readily filtered down to 0.2 μm. Fraction IV-4, on the other hand, produces an equally small but loose and intensely coloured pellet, and although the supernatant can be readily filtered down to 0.8 μm, filtration through 0.45 and 0.2 μm membranes is much more difficult. The preparations are sterilized for use by filtration at 0.2 μm. The yield is about 240 g of protein per kg of paste. The product is used at 5% (v/v) of the final culture medium. Both fractions IV-1 and IV-4 processed in this way have been found to support the growth of hybridomas producing monoclonal antibodies as effectively as foetal calf serum [1].

The product described by Ng and Dobkin is an extract of a fraction IV collected at pH 6.05 and 40% (v/v) ethanol which is broadly similar to fraction IV-4 of the Protein Fractionation Centre. To evaluate the effect of the different procedures on the composition of the products, fractions IV-1 and IV-4 produced at the Protein Fractionation Centre were extracted using both the procedure of Ng and Dobkin and that of MacLeod. The products were analysed by polyacrylamide gradient gel electrophoresis (Fig. 3). The only clear difference between the products of the two methods is the presence of a greater amount of low molecular weight material in the products of MacLeod particularly from fraction IV-4. There is no evidence of a general reduction in the content of high molecular weight material in the product of the Ng and Dobkin process, although this does not preclude the selective loss of specific components of the high molecular weight material. The products of both fraction IV-1 and fraction IV-4 produced by either process were found to be capable of sustaining growth and monoclonal antibody production of murine hybridoma cells.

The conflicting experiences reported by Ng and Dobkin and by MacLeod are not easily reconciled. The results of MacLeod suggest that there is not necessarily a significant amount of general cell growth inhibitory material to the found in fractions IV-1 or IV-4. The various forms of fraction IV are all heterogeneous mixtures enriched for α- and β-globulins but containing all the other classes of plasma protein notably albumin. Not all these proteins are in their native form however. The lipids constituting part of the lipoproteins, for instance, are substantially redistributed [44] so that growth inhibition in a particular case could conceivably be attributed to presence of an active cell growth inhibitory factor, to loss of specific lipids that are nutritionally essential, or to the presence of free lipids acting as detergents [29]. The ability of a particular batch of fraction IV material to support cell growth may reflect the quality of the source plasma and the way in which it has been handled. In particular the endotoxin resulting from bacterial contamination has been shown to have an adverse effect on some cells in vitro [46] and this can be detected in fraction IV. Thus it is possible that different manufacturers drawing on plasma from different sources, using different versions of the Cohn process and handling their in-process material in different ways, may produce fractions IV that have different capacities for supporting cell growth.

4.2 Protein Supplements from Sources other than Fraction IV

Many proteins have been found to have a stimulating effect on animal cells being cultured in vitro and some of these are present in plasma but are not recovered in the Cohn fraction IV discussed above. In particular fibronectin has been identified as being important in promoting the adhesion and spreading of cells that need to attach to a surface before they can grow and divide [47]. Fibronectin is found in fraction 1 of the Cohn process or it may be recovered during the production of coagulation factor VIII from cryo-precipitate [48]. The major protein in these fractions is fibrinogen but the preparation can be defibrinated by heating at 56 °C. The fibrinogen will denature and precipitate at this temperature but the fibronectin, having a greater thermal stability, remains in solution [21].

Proteins that have been found to stimulate cells in vitro can be recovered from fractions other than fraction IV. Thrombin has been shown to be mitogenic for some cell types [49] and can be purified from serum fraction III [50], although a simpler approach would be to recalcify prothrombin complex concentrates prepared by ion exchange chromatography of plasma cryo-supernatant [51]. α_1-acid glycoprotein has also been found to stimulate some types of cells. This protein can be recovered from the supernatant of fraction IV-4 + V by batch-adsorption onto DEAE ion exchange medium, or by extraction from Cohn fraction VI [52].

To accommodate emergencies in routine blood bank operation it is necessary to maintain stocks of components at a level above that required by immediate demand. Consequently a proportion of the stocks will reach the end of their shelf-life without being used. These out-dated components have been used successfully as cell culture medium supplements. Zolg et al. [20] used plasma collected from out-dated whole blood and defibrinated by adsorption onto kaolin to support malarial parasite culture in human red cells. Duffy et al. [19] recalcified similar plasma to form a clot and found that the supernatant was comparable with conventional pooled human serum in its ability to support mixed lymphocyte culture. Schwartz et al. [53] used clonal growth of normal human skin fibroblasts to evaluate serum produced by recalcifying out-dated platelet concentrates. The product was found to support formation of a similar number of colonies as foetal cell serum but the colonies grew more rapidly in human platelet serum supplemented medium.

5 Problems Associated with the Use of Plasma Protein Fractions

Use of human plasma protein fractions in animal cell culture does pose some problems. The fractions are far from pure and this may lead to difficulties if extraneous protein interferes with the cell product. An example of this is the presence of immunoglobulin in fraction IV-1 which may interfere with recovery of monoclonal antibodies if a pure product is required, for instance in construction of immuno-affinity systems.

The major problem associated with the use of plasma fractions is the possibility of contamination with infectious viruses. However the recent increase in the danger of infection arising from use of blood products has accelerated developments of processes for inactivation of infectious, and in particular viral, contaminants. Inactivation processes may involve heating the product either as a freeze-dried powder or in solution, irradiation [54] or the disruption of lipid enveloped viruses using detergents [55].

The production of proteins for clinical use by animal cell culture is complicated by the fact that the cells themselves may be producing virus and thus contaminating the product [56]. In many cases the identities of such viruses are unknown and their presence must be deduced from detection of enzyme activity such as reverse transcriptase. Thus there is a strong case for subjecting all animal cell products to the most vigorous inactivation procedure that the product will tolerate and this will serve also to inactivate any virus derived from the protein supplement added to the medium.

6 Conclusion

The prospects for extensive exploitation of Cohn cold-ethanol fractions of human plasma proteins as components of animal cell culture media are at best uncertain. The fractions most readily available, those that are discarded at present, are complex mixtures of proteins and each includes most of the identifiable plasma proteins although in changed proportions from whole plasma. The fractions produced by a single manufacturer show batch-to-batch variation in composition and may differ widely between manufacturers. The complexity of the plasma protein mixture obtained from these fractions and the variability of its composition re-introduces some of the long-standing criticisms of the use of whole serum and is compounded by the possibility that infectious viruses may contaminate human blood products.

The advantages of utilising whole plasma protein fractions include the financial consideration that these fractions are already routinely being produced and that resolution, clarification, sterilisation and bottling may not add greatly to the overall cost of the operation. A further consideration is that for therapeutic applications there could be merit in having the cell culture derived protein (e.g. monoclonal antibody) formulated in a mixture of human plasma proteins so that the final product resembles existing plasma protein preparations for clinical use. This may serve to reduce the specific immune response seen after the administration of homogeneous proteins, for instance in the development of anti-idiotype antibodies to pure monoclonal antibodies [57]. Consequently there may be no particular disadvantage to producing these proteins in a culture medium supplemented with plasma protein fractions. It would not be necessary to purify the product extensively, the only provisions being to ensure that protein carried through from the culture medium is not damaged, for instance by generation of aggregates or fragments, does not itself damage the product for instance by activation of proteases [11], and that an effective process is available to ensure that any infective contamination is inactivated in the final product.

For some applications as in the production of immuno-affinity purification reagents, it will not be acceptable to have the cell product substantially contaminated with protein carried over from the culture medium. Consequently it will be necessary to grow the cells in a serum-free medium containing very low levels of protein to minimise the degree of purification required. In this case the Cohn fractions can be regarded as ready sources for the preparation of important proteins including transferrin, albumin, insulin-like growth factors, caeruloplasmin, α2-macroglobulin, α1-antiprotease, thrombin and fibronectin. Apart from reducing the protein content of the medium and thus the problems of downstream processing, the use of purified proteins allows the level of each protein to be optimised for a given cell line which may lead to more effective use of the plasma fractions available. However production of purified proteins will require substantial additional investment beyond recovery of the Cohn Fraction.

The human plasma protein fractions which are by-products of routine plasma fractionation represent a valuable resource for animal, especially human, cell culture that should not be disregarded. In particular they could represent an important component in the development of safe and cost effective therapeutic products from cell culture systems.

7 Acknowledgement

The author is grateful to colleagues in all departments of the Protein Fractionation Centre for their contributions to the preparation of this review and to some of the work described, especially Dr P. R. Foster for his helpful comments during the preparation of the text.

8 References

1. MacLeod, A. J., Thomson, M. B.: Develop. Biol. Stand. *60*, 55 (1985)
2. Lambert, K. G., Birch, J. R.: Cell Growth Media, In: Animal Cell Biotechnology, Vol. 1 (Spier, R. E., Griffiths, J. B., eds.), p. 85, London: Academic Press 1985
3. Tytell, A. A., Scattergood, E., Field, A. K.: U.S. Pat. 4,198,479 (1980)
4. Barnes, D., Sato, G. H.: Analyt. Biochem. *102*, 255 (1980)
5. Waymouth, C.: Preparation and Use of Serum-free Culture Media, In: Cell Culture Methods for Molecular and Cell Biology, Vol. 1 (Barnes, D. W., Sirbasku, D. A., Sato, G. H., eds.), p. 23, New York: Alan R. Liss 1984
6. Griffiths, B.: Trends in Biotechnology *4*, 268 (1986)
7. Heide, K., Haupt, H., Schwick, H. G.: Plasma Protein Fractionation, In: The Plasma Proteins, Vol. 3 (Putnam, F. W., ed.), p. 545, New York: Academic Press 1977
8. Ogston, D., Bennet, B. (eds.): Haemostasis: Biochemistry, Physiology and Pathology. London: John Wiley 1977
9. Heldin, C.-H., Wasteson, A., Westermark, B.: Mol. Cell. Endocrinol. *39*, 169 (1985)
10. Heldin, C.-H. et al.: J. Cell Sci., Suppl. *3*, 65 (1985)
11. Heldin, C.-H. et al.: Nature *319*, 511 (1986)
12. Balk, S. D. et al.: Proc. Nat. Acad. Sci. USA *78*, 5656 (1981)
13. Weinstein, R. et al.: J. Cell. Phys. *110*, 23 (1982)
14. Dafgard, E. et al.: J. Cell Sci., Suppl. *3*, 53 (1985)
15. Hintz, R. L., Liu, E.: Serum Forms of Insulin-like Growth Factors and their Carrier Proteins, In: Insulin-like Growth Factors/Somatomedins (Spencer, E. M., ed.), p. 133, Berlin: Walter de Gruyter 1983
16. Scher, C. D. et al.: J. Cell. Phys. *97*, 371 (1978)
17. Stiles, C. D. et al.: Proc. Nat. Acad. Sci. USA *76*, 1279 (1979)
18. Conover, C. A. et al.: J. Cell. Phys. *116*, 191 (1983)
19. Duffy, B. F., Oldfather, J. W., Rodey, G. E.: J. Immunol. Meths. *79*, 223 (1985)
20. Zolg, J. W. et al.: J. Parasitol. *68*, 1072 (1982)
21. Ingham, K. C. et al.: J. Biol. Chem. *259*, 11901 (1984)
22. Cohn, E. J. et al.: J. Am. Chem. Soc. *68*, 459 (1946)
23. Kistler, P., Friedli, H.: Ethanol Precipitation, In: Methods of Plasma Protein Fractionation (Curling, J. M., ed.), p. 3, London: Academic Press 1980

24. Foster, P. R., Watt, J. G.: The CSVM Fractionation Process, In: Methods of Plasma Protein Fractionation (Curling, J. M., ed.), p. 17, London: Academic Press 1980
25. Laemmli, U. K.: Nature *227*, 680 (1970)
26. Bazeley, P. L. et al.: Proc. Soc. Exp. Biol. Med. *87*, 420 (1954)
27. Sandford, K. K. et al.: J. Nat. Cancer. P In. *16*, 789 (1955)
28. Chang, R. S. et al.: Proc. Soc. Exp. Biol. Med. *102*, 213 (1959)
29. De Luca, C., Carruthers, C., Tritsch, G. L.: Exp. Cell Res. *43*, 451 (1966)
30. De Luca, C., Rakowski, P. F., Tritsch, G. L.: Biochim. Biophys. Acta. *86*, 346 (1964)
31. Pirt, S. J., Lambert, K.: Develop. Biol. Stand. *37*, 63 (1977)
32. Melnick, J. L., Wallis, C.: ibid. *37*, 77 (1977)
33. Salomon, D. S. et al.: α2-Macroglobulin, a Contaminant of Commercially Prepared Pedersen Fetuin: Isolation, Characterisation and Biological Activity, In: Cell Culture Methods for Molecular and Cell Biology, Vol. 3 (Barnes, D. W., Sirbasku, D. A., Sato, G. H., eds.), p. 125, New York: Alan R. Liss 1984
34. Spieker-Polet, H., Cruise, S. A., Polet, H.: Cell. Immunol. *44*, 144 (1979)
35. Iscove, N. N., Melchers, F.: J. Exp. Med. *147*, 923 (1978)
36. Blaker, G. J., Birch, J. R., Pirt, S. J.: J. Cell Sci. *9*, 529 (1971)
37. Uthne, K.: Acta Endocrinol. *73*, suppl. 175 (1973)
38. MacLeod, A. J.: Nature *285*, 136 (1980)
39. Weinstein, R. et al.: J. Cell. Phys. *110*, 23 (1982)
40. Blum, W. F., Ranke, M. B., Bierich, J. R.: Acta Endocrinol. *111*, 271 (1986)
41. Ng, P. K., Dobkin, M. B.: U.S. Pat. 4,452,893 (1984)
42. MacLeod, A. J.: U.K. Pat. Appl. 8,430,079 (1985)
43. Ito, I. et al.: J. Cell. Phys. *113*, 1 (1982)
44. Pennell, R. B.: Fractionation and Isolation of Purified Components by Precipitation Methods, In: The Plasma Proteins Vol. 1 (Putnam, F. W., ed.), p. 9, New York: Academic Press, 1960
45. Harrington, W. N., Godman, G. C.: Proc. Nat. Acad. Sci. USA *77*, 423 (1980)
46. Price, P. J., Gregory, E. A.: In Vitro *18*, 576 (1982)
47. Yamada, K. M. et al.: Biochemistry *16*, 5552 (1977)
48. Horowitz, B. et al.: Transfusion *24*, 357 (1984)
49. Carney, D. H., Glenn, K. C., Cunningham, D. D.: J. Cell. Phys. *95*, 13 (1978)
50. Fenton, J. W. et al.: J. Biol. Chem. *252*, 3587 (1977)
51. Brummelhius, H. G. J.: Preparation of the Prothrombin Complex, In: Methods of Plasma Protein Fractionation (Curling, E. J., ed.), p. 117, London: Academic Press 1980
52. Maeda, H. et al.: Proc. Soc. Exp. Biol. Med. *163*, 223 (1980)
53. Schwartz, K. A. et al.: Am. J. Hematol. *17*, 23 (1984)
54. Horowitz, B. et al.: Transfusion *25*, 523 (1985)
55. Horowitz, B. et al.: ibid. *25*, 516 (1985)
56. Lubiniecki, A. S., May, L. H.: Develop. Biol. Stand. *60*, 123 (1985)
57. Jaffers, G. J. et al.: Transplantation *41*, 572 (1986)

Note Added in Proof

The use of Cohn fraction IV-1 material prepared by the method of MacLeod to sustain MDCK cells growing on microcarriers has been reported. The fraction IV-1 material was found to give a greater yield of cells than either 10% foetal calf serum supplement or serum free media. Sayer, T. E., Butler, M. and MacLeod, A. J., Proceedings of the European Society for Animal Cell Technology 8th Meeting, Tiberias, Israel, 1987, to be published in Developments in Biological Standardization.

Industrial Scale Production of β-Interferon

M. Morandi and A. Valeri
Biological Research and Development Sclavo S.p.A., Via Fiorentina 1, 53100 Siena/Italy

The first part of the survey is dedicated to the current status of the large scale production of β-IFN, intended for extensive and significant clinical trials. In particular, the methods employed for the growth of human foreskin fibroblasts and the induction and purification of the active molecule are reviewed.

The second part is dedicated to the detailed description of the only plant operating in Italy, producing clinical grade IFN. Special emphasis is given to the control methodologies adopted for the registration of the product.

Finally, some consideration is given to the present state of the art and the future trends in the field of β-IFN production.

1 Introduction

Studies carried out by Youngner et al. [1] and further developed by Vilcek et al. [2,3] and Tan et al. [4] led to the formulation of a superinduction schedule of human diploid fibroblasts, for the production of β-IFN.

The practical application of these studies led to the preparation of this IFN in quantities that allowed its chemical-physical characterization and its differentiation from other IFNs produced by other types of cells [5-7].

The selection of human fibroblasts as choice substrate for production, made clear a series of advantages and disadvantages associated with the large scale development and clinical use of β-IFN.

Safety is undoubtedly the main advantage in using these cellular systems, and in fact human diploid fibroblasts have long been used and considered safe for the production of viral vaccines [8] and are therefore well characterized regarding karyotype and adventitious agents. For other possible cell substrates, such as primary cultures (for example lymphocytes) or continuous cell lines, it is impossible to obtain a similar level of confidence. In fact, primary cells are sometimes contaminated by known and unknown viruses (Hepatitis B, non-A non-B, retroviruses, etc.) and continuous cells are transformed by important karyological modifications or by extraneous genetic material (i.e. viral integrated genes), potentially harmful for human beings. However, an industrial scale IFN production based on fibroblasts requires expensive facilities and advanced technologies because of their anchorage dependence and their finite life span.

Great hopes of overcoming all the drawbacks connected with the massive production of IFN arose with the development of genetical engineering techniques. β-IFN was cloned and expressed in *Escherichia coli* [9-11] with good production yields, but the molecules obtained with this system were not glycosilated [10] as opposed to natural IFN. This lack of glycosilation caused serious stability problems during purification, and strongly limited clinical testing; this fact is confirmed by the absence of literature regarding clinical trials using recombinant β-IFN.

In order to obtain a product as close as possible to the natural one, the human β-IFN gene was successfully cloned and expressed in eukaryotic cells, such as yeast or mammalian cells [12-18], in which the molecule can be obtained as a glycoprotein. However, in the yeast system, β-IFN resulted toxic for the expression vector itself (authors unpublished results) and the mammalian cell approach is still being developed.

Furthermore, many important aspects regarding the glycosilation of recombinant molecules are still unknown [19] and may be relevant during their clinical use in human beings. Consequently, the natural product is still very important, especially if one considers the acceptable tolerability and the good efficacy shown in many successful clinical trials [20-22] and Sclavo (unpublished data).

2 Background

The technological limitations caused by the anchorage dependence of human diploid fibroblasts and the expenses involved in the organization required the development and control of the production of β-IFN on industrial scale, thus limiting the number of companies able to obtain clinical grade β-IFN for extensive testing in human beings.

Van Damme and Billiau [23] mention twelve research groups that a few years ago were involved in this production. Today only Bioferon (Rentschler) in Germany, Toray in Japan and Inter-Yeda/Serono in Israel/Italy have completely developed the process, up to clinical trials and registration. Our italian group is to be added to this list as we have developed, since 1980, a production process that has permitted the registration of our β-IFN in Italy, following the successful completion of clinical trials against several viral diseases.

2.1 Growth of Human Fibroblast Cells in Large Scale

Among the many techniques tried for the growth of anchorage dependent cells [24], we will describe only those proved suitable for large scale human fibroblast expansion and therefore adopted by industries or institutions interested in hu β-IFN production for extensive clinical use.

Roller Bottles

This can be considered the simplest technological approach and is based on the linear expansion of a large quantity of small batch reactors [25]. Easily suitable for a vaccine production, in which relatively few bottles are required, many difficulties are encountered in a very large scaling up such as that needed for β-IFN production. It is extremely complex to handle hundreds or thousands of these batch reactors simultaneously: even the simplest change of medium is highly laborintensive and has to be done very accurately under strictly sterile conditions.

Although currently employed as the initial step of production, the space needed, the cost of equipment and labor make of this technique one of the most expensive and outdated ones, despite all attempts of automation.

Multitrays

This approach, which essentially reduces the drawbacks encountered in the roller bottle system, is based on the use of single batch reactors called "multitrays" [26]. The most widely used are multitray units [27] composed of 11 or more polystyrene trays assembled to form 10 or more culture chambers. The trays are connected and intersected by two channels that allow the trays to be filled with exactly the same volumes of medium following the principle of communicating vessels. After filling, which occurs in a vertical position, the multitrays are changed into a horizontal working position rendering all the trays completely independent one from the other.

The multitrays are difficult to manipulate due to their very large size and frangibility. Their use in processes such as the production of IFN, that involves frequent washings and addition of new components, appears feasible only by a complete automation of the system. Anyhow, as with the roller bottles technology, any expansion implies a proportional increase of space and labor.

Multisurface Propagators

This system, originally developed for the production of viral vaccines [28,29], was adapted and successfully modified by G. Mann (London School of Hygiene and Tropical Medicine, personal communication) for the production of Hu β-IFN in USA. The culture vessels consist of an open sided box fitted with inlet and outlet ports containing glass plates separated and sealed by teflon strips. Vessels of any desired growth surface area in the range of $\geqq 0.01$ to $\leqq 10\ m^2$ can be constructed, ensuring uniform operational characteristics. Vessels are connected through a pump to the medium reservoirs and diffusion manifold located adjacent to the open channels ensure an even distribution of the medium between the plates. It is possible to obtain high cellular densities by continuous perfusion of the plates, since the cell saturation density level is directlv related to the volume of the medium used.

The scaling up of such a system presents some limitations mainly during the trypsinisation and the recovery of viable cells. Other drawbacks are the difficulties of visualization of the monolayers and the impossibility of controlling the major growth parameters (pH, pO_2) inside the culture vessels.

Microcarriers

In these last years the microcarrier technique has been widely developed for the large scale growth of anchorage dependent cells and it is based on the principle devised by Van Wezel [30], of growing the cells on very small (150–200 μm diameter) and low density (1.03) spherical carriers in order to allow suspension in liquid media with light stirring. This technique is characterized by a wide growth surface in limited volumes of medium and by the possibility of working with homogeneous and submerse cultures with the advantage of automatic control of the main growth parameters (T, pO_2, pH, pCO_2, etc.) and direct continuous monitoring of the cultures.

The microcarrier system usually allows an easy scaling up without a proportional increase of personnel and space. Thanks to this technology, Toray in Japan developed the world's largest facility for the production of Hu β-IFN from fibroblasts, with culture volumes of up to 4000 liters.

Many other systems for large scale cultivation of mammalian cells have been proposed, such as air-lift bioreactors [31], ceramic cartridges [32] or hollow fibers [33], but up to now they are not used in β-IFN production and have generally found greater employment in the growth of lymphoblastoid cell lines or hybridomas.

2.2 Interferon Induction

No virus is used to induce fibroblasts for the production of β-IFN, as opposed to lymphocyte stimulation (usually by Sendai virus) for the production of α-IFN. A synthetic double-stranded RNA —Poli(rI)Poly(rC)— is employed, thus simulating a viral infection which induces the cell to synthesize β-IFN molecules.

All manufacturers of natural β-IFN use this procedure, improved by superinduction where the use of a translation inhibitor, cycloheximide, followed by a transcription inhibitor, actinomycin D, allows an increase of the final IFN yield, protracting its synthesis [2–4].

2.3 Purification

The acquisition of large amounts of cells for the production of crude β-IFN is certainly of paramount importance, but the purification process is of equal importance because it must lead to a safe product with an economically acceptable yield.

Many methods have been proposed for the purification of IFNs [34–41], but most of them are suitable only for laboratory use, either due to the presence of chemical substances of controversial acceptability for human administration or because, at the present stage of our knowledge, they could hardly be transferred to an industrial scale owing to their technological inadequacy, high costs and/or low yield.

It is certainly not easy to know all the details regarding the purification systems used by the manufacturers of Hu β-IFN, nonetheless it is possible to try to illustrate the most significant techniques that have led to the preparation of injectable clinical grade Hu β-IFN.

According to the suggestions of the USA Office of Biologics [42], Hu β-IFN produced in homologous cells, such as human diploid fibroblasts, should be used for clinical trials only if its specific activity is higher than 10^6 IU per mg of protein. This low level of purity, which is far below either the molecular homogeneity ($1 \div 2 \times 10^9$ IU per mg of protein (43)) or the level usually required for recombinant Hu IFN expressed in heterologous vectors (42), is accepted only for human-derived preparations because of the homologous nature of the foreign contaminating proteins (self proteins). Therefore, when dealing with human natural β-IFN, the purification processes are relatively simpler, with a good yield. Generally, it is also possible to eliminate drastic denaturating conditions thus avoiding the risk of structure modifications of the glycoprotein.

Controlled Pore Glass

CPG chromatography is used as a first, and sometimes as the only step able to obtain clinical grade β-IFN. It is based on the hydrophobic properties of the β-IFN molecules that can be adsorbed on glass at neutral pH and subsequently eluted in acid conditions.

Edy et al. [44] report the use of CPG to obtain Hu β-IFN 90-fold purified with yields varying between 75% and 100%.

This technique was partially modified by Van Damme and Billiau [23] and further improved by the same group, adding a second step consisting of Zinc-chelate chromatography [45,46]. With this modification, an intermediate step of dialysis is necessary. The combined methods led to an overall recovery of about 70% with specific activity of up to 2×10^9 units per mg of protein.

Zinc-Chelate Chromatography

As already mentioned, this technique is widely used and generally employed with partially purified β-IFN. This method is based on the observation that Zn^{2+}-imino-diacetate-sepharose complex is able to retain β-IFN at a low salt concentration and neutral pH. The elution is performed at high ionic concentration and low pH, with good recovery and a specific activity of about 10^8 units per mg of protein [47]. For the same purification purposes it is also possible to use other chelates such as Co^{2+}, Ni^{2+} and Cu^{2+}.

Agarose Immobilized Concanavalin-A Chromatography

The β-IFN molecule is a glycoprotein and, as such, its carbohydrate moiety has affinity with lectins. This sugar-lectin interaction was used to develop an affinity chromatography purification technique using Con-A as lectin [25]. The bound IFN is displaced from the matrix with 50% ethylene glycol in the presence of 0.1 M methyl-α-D-mannopyranoside. Under these conditions a 2000-fold purification was obtained with specific activity of $1 \div 5 \times 10^7$ units per mg of protein with average yields of approximately 60%.

This first step of lectin chromatography is completed by a further treatment of the partially purified IFN in a column of phenyl-sepharose in order to avoid the presence of Con-A residues (mitogen) and to increase purity. β-IFN binds to the phenyl-sepharose matrix by hydrophobic interaction and it is displaced again with 75% EG.

Concentrated preparations with specific activity of up to 5×10^7 units per mg of protein are obtained with almost 100% recovery.

Molecular Sieve Chromatography

This technique has many applications for protein separation purposes and it has also been applied as the main purification step for Hu β-IFN [27]. Using Sephadex G75, a gel with a lower exclusion limit of 50 ÷ 100 kDa, IFN was purified up to a specific activity of 10^6 IU per mg of protein.

The inconvenience of gel chromatography is undoubtedly the high dilution which IFN undergoes during this procedure, therefore it is necessary to strongly concentrate both the crude starting material and the purified eluted IFN. The former condition was obtained by precipitating the crude material with perchloric acid and ammonium sulphate, and subsequently dissolving the precipitate with sodium hydroxide. Following gel chromatography, the fraction selected with the highest content of β-IFN is concentrated by ultrafiltration up to the level of 2×10^6 IU ml^{-1}.

3 Large Scale Production of Natural β-Interferon in Italy

We selected the production of Hu β-IFN from diploid cells (human foreskin fibroblasts) for two main reasons: first, our company already had an advanced research group in the field of large scale cell cultures and therefore could undertake a program for the production of a biomolecule starting from human diploid cells, and second, we considered this type of IFN the best possible choice with regards to safety for human use.

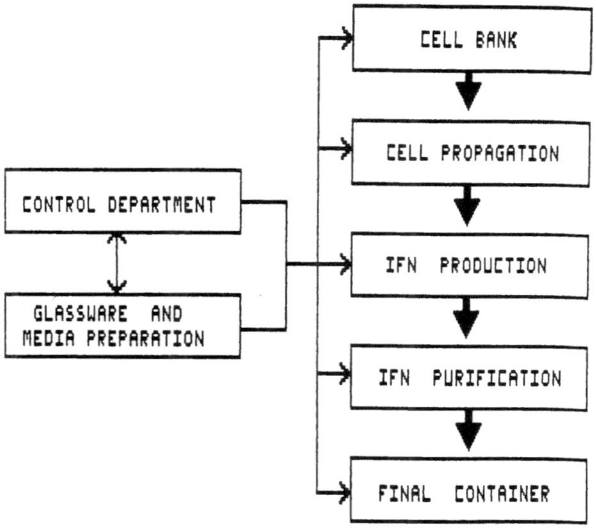

Fig. 1. Main sections of the production plant

An outline of the organization of the whole production plant is shown in Fig. 1. It refers to a pilot plant with a weekly production capacity of 100 liters of crude IFN, obtained intwo 50 L bioreactors. This level of production is sufficient to treat thousands of patients with viral diseases or hundreds of cancer patients (who require higher dosages and longer periods of treatment) each year and was therefore considered adequate to start meaningful clinical trials at acceptable costs and on a large enough scale to collect technical data at an industrial level for future expansion.

The whole plant, excluding the glassware and media preparation section, was built in an area provided with an independent air conditioning system that generates and maintains a positive pressure with respect to the atmospheric pressure inside the laboratories. All the air outlet ports of this conditioning system are provided with absolute class 100 HEPA filters, i.e. the particles count inside the plant is not exceeding a total of 3.5 particles per liter.

3.1 Cell Bank

Cell strains derived from neonatal foreskin and selected for IFN production are preserved in liquid nitrogen as master seed at PDL 18. Following accurate controls of the master seed (see Sect. 3.6), working seeds are periodically prepared at PDL 24, as glass sealed vials, each one containing about 30×10^6 cells. The working seeds constitute the starting point for the production, which takes place in cell cultures at a maximum PDL of 33.

3.2 Cell Propagation

Based on our previous research experiences on the economy and practicality of the system, our choice was to use the microcarrier technique in submerse cultures.

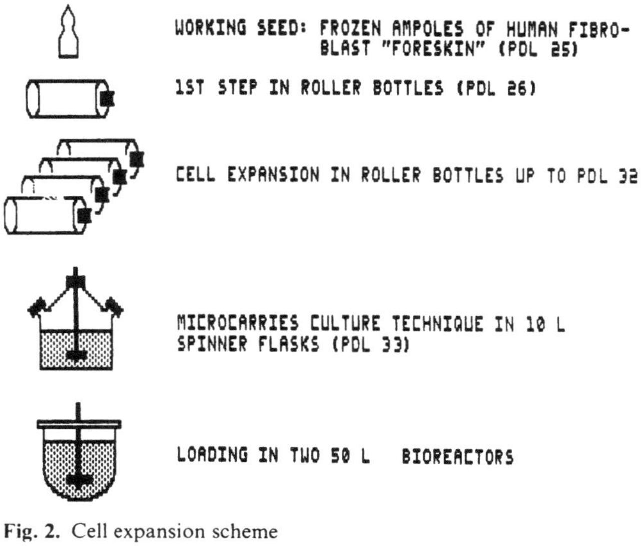

Fig. 2. Cell expansion scheme

The cell growth expansion takes about 40 days (6 weeks) during which the cells are first cultivated in RB, are transferred into SF and then into bioreactors (Fig. 2). DMEM supplemented with 10% FBS is the medium employed for the RB step while for the microcarrier cultures (SF and bioreactors), a 5% FBS reduction in the same medium is operated. The cells grown in RB are weekly split at a ratio of 1:3 ÷ 1:4, until obtaining enough cells to seed 24 ÷ 30 SF (4 L volume) at an incubation density of 3×10^5 cell ml^{-1} with a microcarrier concentration of 4 g L^{-1}. After four days, the spent medium is discarded and the cultures are collected and transferred with fresh growth medium in two 50 L bioreactors. These glass bioreactors possess a cylindrical geometric configuration, with a height-to-diameter ratio of 2:1, in order to obtain the most efficient gas-liquid mass transfer and a good homogeneity of the cultures with a minimum stirring speed. The pH, pO_2 and temperature are measured by probes directly immersed in the cultures, and maintained at the set values by conventional analog controls. The pH (7.2) is controlled by CO_2 addition to the inlet gas stream (acid correction) or by $NaHCO_3$ injection in the cultures (basic correction). The pO_2 is maintained at 50% air saturation, increasing or reducing the O_2 content in the gas mixture flowing in the bioreactor's head space.

After 48 h, the cells reach the maximum saturation density (1 ÷ 1.2×10^6 cells per ml^{-1}) and are finally ready to be superinduced.

3.3 Interferon Induction

The cell culture is superinduced for the production of Hu β-IFN according to the classical schedule [48], with minor variations. Our process begins by washing the cultures twice with Earle solution and by priming the fibroblasts on the microcarriers with 50 ÷ 100 IU ml^{-1} of Hu β-IFN. After 15 h, 50 μg of Poly(rI)Poly(rC) and 10 μg ml^{-1} of Cycloheximide are added, followed 4 h later by 1 μg ml^{-1} of Actinomycin D. The culture is finally washed several times with Earle solution, and DMEM with 0.025% human albumin is added for the IFN production phase, which occurs at 30 °C for about 36 h. Subsequently, the supernatant containing the crude β-IFN is separated from the microcarriers and harvested in a sterile stainless steel reservoir.

3.4 Purification

The possibility of obtaining high purification levels and very good yields in large scale operations with only one step, and considering that the matrix employed guarantees the complete absence of toxic or dangerous substances for man, convinced us to adopt and attempt to improve the CPG chromatography technique.

The clarified crude IFN is loaded overnight through a CPG column in a cold room at +4 °C (flow-rate at 4 L h^{-1}) with a completely automatic system, equipped with a peristaltic pump. On the following morning, after extensive washings of the matrix, the elution of purified β-IFN is obtained with a step gradient of acetic acid.

With this procedure, up to 1000-fold purification is generally achieved, with about 70% yield (Table 1). The fractions containing the purified Hu β-IFN are ready for the final processing without any further treatment and may be stored at +4 °C up to 6 months without loss of activity. It is our experience that freezing the purified

Table 1. CPG concentration and purification of Hu β IFN

Crude				CPG purified			
Prod. cycles (n.)	Vol. (L)	IFN yield ($IU\ ml^{-1}$)[a]	Spec. activ. ($IU\ mg^{-1}$ prot.)	Vol. (L)	IFN yield ($IU\ ml^{-1}$)[a]	Spec. activ. ($IU\ mg^{-1}$ prot.)	Recovery (%)[b]
20	70	2.0×10^4	1.3×10^4	2.0	49.0×10^4	1.2×10^7	70
150	75	1.9×10^4	1.2×10^4	1.9	48.7×10^4	1.1×10^7	65
210	65	3.1×10^4	1.5×10^4	2.2	66.8×10^4	1.4×10^7	73
Average[c]	72	2.4×10^4	1.4×10^4	2.1	56.0×10^4	1.3×10^7	69

[a] IFN biological activity obtained by cytopathic effect reduction assay [54] and expressed as International Units using the National Institutes of Health (Bethesda, MD, USA) Reference Preparation G-023-902-527;
[b] percentage of initial amount of crude IFN;
[c] average data of the total production (1981—1986)

material at —80 °C and subsequently thawing it, can cause a remarkable loss of biological activity.

3.5 Final Processing

The selected fractions which pass the necessary quality controls are mixed and, after the addition of human albumin and mannitol, the pool is sterilized by filtration and properly distributed in glass vials containing $1\text{–}3 \times 10^6$ IU per vial.

The filtration is performed through a polysulfone membrane previously treated with a human albumin solution in order to avoid adsorption of IFN on the membrane itself. The vials are lyophilized in 48 h, sealed and stored at —20 °C before distribution, since at this temperature the stability of the product, calculated by the accelerated

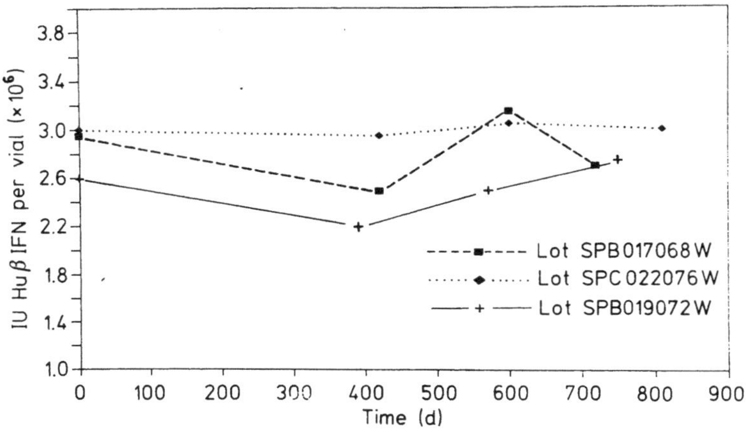

Fig. 3. Natural degradation rate of three representative lots of lyophilized Hu β-IFN stored at —20 °C

Table 2a and b. Set of controls for Hu β interferon production

a

Master and working seeds	Cell cultures	Crude IFN	Purified IFN
Kariology	Sterility	Sterility	Sterility
	— Bacterial and mycotic	— Bacterial and mycotic	— Bacterial and mycotic
Life-span	— Mycoplasma	Antiviral activity (IU ml^{-1})	Antiviral activity (IU ml^{-1})
Oncogenicity	LDH[b]	Protein content (mg ml^{-1})	Protein content (mg ml^{-1})
— Hamster	Limulus test	Limulus test	
— Nude mice	Adventitious agents		
Sterility			
— Bacterial and mycotic			
— Mycoplasma			
Adventitious agents (cell cultures)			
— Monkey kidney cells (cercopithecus)			
— Rabbit kidney cells			
— Chicken embryo fibroblasts (SPF[a] eggs)			
— Homologous diploid fibroblasts			
— Heterologous diploid fibroblasts			
Adventitious agents (SPF[a] chicken embryo eggs)			
— Allantoic cavity			
— Yolk sac			
— Choryoallantoic membrane			
Adventitious agents (animals)			
— Adult mice			
— New born mice			
— Guinea pigs			
— Rabbits			

b

Bulk before filtration	Bulk after filtration	Final containers
Sterility	Sterility	Sterility
— Bacterial and mycotic	— Bacterial and mycotic	— Bacterial and mycotic
— Mycoplasma	Antiviral activity (IU ml^{-1})	Antiviral activity (IU per vial)
Antiviral activity (IU ml^{-1})	Protein content (mg ml^{-1})	Pyrogens (rabbits)
Protein content (mg ml^{-1})		Limulus test
		Toxicity
		— Mice
		— Rats
		Safety tests
		— Guinea pigs
		— Mice
		Specific activity (IU mg^{-1} of protein)
		HSA[c] identity
		IFN identification
		— Specific antiserum
		— Trypsin sensitivity
		— pH 2 stability
		— Ultracentrifuge (no sedimentation)
		Labels control
		Vials leakage
		Residual humidity
		Solubility

[a] SPF = Specific pathogen free
[b] Cell identification by lactic dehydrogenase (LDH) isoenzyme electrophoresis
[c] HSA = Human serum albumin

degradation rate test [49], is virtually unlimited. These data are furthermore supported by the results obtained with the natural degradation test (Fig. 3).

3.6 Controls

The recent discovery of biologically active molecules characterized by high therapeutic potential (interleukins, interferons, tissue plasminogen activator, tumor necrosis factor, epidermal and nerve growth factors, etc.), has stressed the problem of proper and effective controls to be performed both on the substrates of production and on the product itself, in view of possible human clinical trials. In fact, in such a new field, specific regulations are lacking and it is necessary to rely on the existing rules which are already in use, as for instance, in the production of viral vaccines, where cellular substrates similar to those used for the production of biomolecules are adopted [50–52].

Therefore it appears quite logical to apply the existing regulations to the controls of human natural β-IFN, integrating them with more specific techniques regarding the biological activity of the molecule. The situation appears to be quite different when controlling biomolecules obtained from continuous cell lines or expressed in prokaryotic and eukaryotic vectors by means of genetical engineering techniques which, if accepted for use in human beings, will need *ex novo* regulations.

Further consideration should be given to the fact that viral vaccines, which are administered only a few times during a lifetime, are subject to rigid control regulations. It is therefore only reasonable to expect that a drug like IFN, which must be administered numerous times in high dosages, should adhere to at least the same rigid standards.

In fact, an approach of this kind is suggested also by the major international regulatory organizations [42] and is substantially followed by the groups producing natural β-IFN for clinical trials [25,27].

Finally, it should be emphasized that a biomolecule used as a therapeutic drug should be thoroughly tested in animals from the pharmacological and toxicological point of view, before human administration, much more extensively than a new vaccine.

Based on the above mentioned principles, we developed a panel of controls which is presented in Table 2.

4 Concluding Remarks

The high level of technology and the complexity of the production and controls, account for the fact that only a few groups have developed an industrial plant for natural β-IFN production. The introduction of advanced techniques and a rational organization can only partially solve some technical and economical difficulties of this type of production. In fact, besides the solely technical aspects, the chief limiting factor is the biological impossibility of increasing the current yield of natural fibroblast IFN per cell, which does not exceed 0.04–0.05 IU per cell. The industrial expansion of such a system also includes the risk of not being able to satisfy the future demands of β-IFN.

The organizational problems of our pilot plant, which already employs innovative

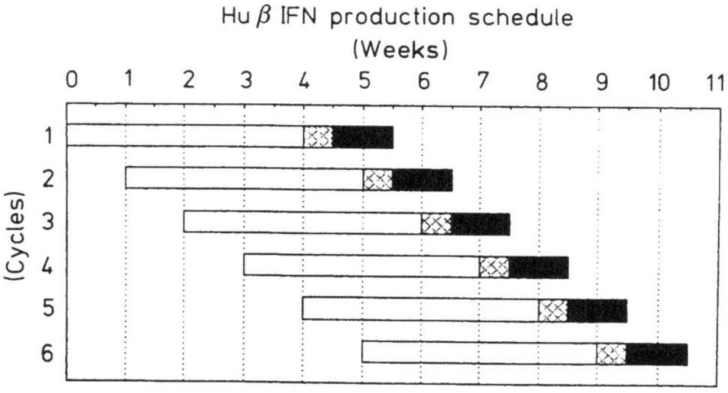

Fig. 4. Overlapping of different working phases: □ roller bottles; ⊠ spinner flasks; ■ agitated vessels (bioreactors) 50 L total volume

Table 3. Total number of culture vessels and related level of contamination during a five year production period

Culture vessels	Inoculated	Contaminated	
Roller bottles	145,000	2,480	1.7
Spinner flasks	6,200	807	13
Bioreactors	520	88	17

technologies, are a good example of this kind of difficulty. Figure 4 shows the complexity and the overlapping of the 5 different working phases which contemporaneously occur on a weekly basis; Table 3 shows the remarkable work load and the risk of product loss due to possible contaminations in the course of this production. Other critical aspects have already been analyzed in a previous study [53].

There is no doubt that the IFN yield obtainable from genetically manipulated mammalian cells can be much higher than that obtained from natural sources: in our laboratories, levels of 0.1 ÷ 0.5 IU per cell per day have been routinely achieved with the major advantage of continuously harvesting the medium from the cell culture bioreactor using perfusion techniques, since recombinant cells can constantly express and secrete Hu IFN molecules. These premises suggest that from the technological point of view, the only possible expansion in a true industrial sense, at an acceptable cost, is the one using recombinant cells. A process based on the use of prokariotic cells would be even more economical, as large volumes can be easily grown and very high levels of expression of IFN achieved (up to 30 ÷ 40% of cell body weight). Nevertheless, the molecule concentration inside the organism and the lack of glycosilation create substantial problems (see Sect. 1), as opposed to the IFN secreted by mammalian cells, which is pratically identical to the natural β-IFN at least with regards to physical and chemical characteristics (authors unpublished data).

In conclusion, the adoption of recombinant DNA technology is necessary to obtain high quantities of clinical grade IFN, at an acceptable cost. This is to satisfy worldwide demand and to provide the possibility of further investigations on potential applications. Nevertheless, the use of genetically manipulated material introduces new problems of primary importance, regarding legal regulations. Guidelines for the production of recombinant molecules are presently being evaluated.

However, the issue of clinical efficacy remains open for discussion and the complete acceptability of recombinant β-IFNs can only be evaluated through an extensive series of clinical trials.

5 List of Abbreviations

Con-A	concanavalin-A
CPG	controlled pore glass
DMEM	Dulbecco's modified Earle medium
EG	ethylen glycol
FBS	fetal calf serum
HEPA	high efficiency particulate air
IFN	interferon
IU	international unit
pCO_2	dissolved carbon hydroxide pressure
pO_2	dissolved oxigen pressure
PDL	population doubling level
RB	roller bottle
SF	spinner flask
T	temperature

6 References

1. Youngner, J. S., Stinebring, W. R., Taube, S. E.: Virology *27*, 541 (1965)
2. Vilcek, J., Rossman, T. G., Varacalli, F.: Nature *222*, 682 (1969)
3. Vilcek, J.: Ann. N.Y. Acad. Sci. *173*, 390 (1970)
4. Tan, Y. H., Armstrong, J. A., Ke, J. H., Ho, M.: Proc. Natl. Acad. Sci. U.S.A. *67*, 464 (1970)
5. Havell, E. R., Vilcek, J.: Antimicro. Ag. Chemother. *2*, 476 (1972)
6. Billiau, A., Joniau, M., De Somer, P.: J. Gen. Virol. *19*, 1 (1973)
7. Committee on Interferon Nomenclature Nature, *286*, 110 (1980)
8. Jacobs, J. P.: Serially-propagated human diploid cell: a synopsis of the present position concerning their use for producing viral vaccines and interferon, in: Develop. Biol. Standard. 42, (ed. Inter. Assoc. Biol. Standard.) p. 13, Basel, S. Karger 1978
9. Derynck, R., Content, J., De Clercq, E., Volckaert, G., Tavernier, J., Devos, R., Fiers, W.: Nature, *285*, 542 (1980)
10. Taniguchi, T., Ohno, S., Takaoka, C.: Expression of the cloned genes for human interferon β 1 in *E. coli* and cultural mouse cells, in: Interferons, UCLA Symposia on molecular and cellular biology. 25 (eds. Merigan, T. and Friedman, R.) p. 15, New York, Academic Press 1982
11. Goeddel, D. V., Shepard, H. M., Yelverton, E., Leung, D., Crea, R.: Nucleic Acids Res. *8*, 4057 (1980)
12. Ohno, S., Taniguchi, T.: Nucleic Acid Res. *10*, 967 (1982)
13. Utsumi, J., Itzuka, M., Kobayashi, S.: J. of Interf. Research *4*, 9 (1984)
14. Jankowsky, W. J., Davey, M. W., O'Malley, J. A., Sulkowsky, E., Carter, A.: J. Virol. *16*, 1124 (1975)
15. Gheysen, D., Fiers, W.: J. Mol. Appl. Genet. *1*, 385 (1982)
16. Cannaani, D., Berg, P.: Proc. Natl. Acad. Sci. U.S.A. *79*, 5166 (1982)
17. McCormik, F., Trahey, M., Innis, M., Dieckmann, B., Ringold, G.: Mol. and Cell Biol. Jan., 166 (1984)

18. Pitha, P. M., Reyes, G. R., Raj, N. B. K., Hayward, G. S.: Expression of human β-interferon gene in heterologous cells, in: Interferons, UCLA Symposia on Molecular and Cellular Biology, 25 (eds. Merigan, T. and Friedman, R.) p. 41, New York, Academy Press 1982
19. Van Brunt, J.: Biotechnology *4*, 835 (1986)
20. Billiau, A., De Somer, P.: Clinical use of interferons in viral infections, in: Interferon and interferon inducers: Clinical application (ed. Stringfellow, D. A.) p. 114, 1980
21. Obert, H. J.: Clinical Trials conducted with beta-interferon. The biology of the Interferon in Germany, in: The Biology of the Interferon System (eds. De Maeyer, E., and Schellekens, H.) p. 469, Amsterdam, New York, Oxford, Elsevier Science Publishers 1983
22. Suzuki, H.: Effect of human fibroblast interferon (Hu IFN β) on Hbc antigen-positive chronic active hepatitis, in: Proceedings of the International Symposium on Interferons (ed. Kiishida, T.) p. 259, Japan Convention Services Inc., Osaka Branch 1983
23. Van Damme, J., Billiau, A.: Large scale production of Human Fibroblast Interferon, in: Methods in Enzymology 78 (ed. Pestka, S.) p. 101, New York, London 1981
24. Spier, R. E.: Adv. Biochem. Eng. *14*, 119 (1980)
25. Leong, S. S., Horoszewicz, J. S.: Production and preparation of human fibroblast interferon or clinical trials, in: Methods in Enzymology 78 (ed. Pestka, S.) p. 87, New York, London 1981
26. Skoda, R., Pakos, V., Hörmann, A., Späth, O., Johansson, A.: Communicating Vessel Systems for Mass Cell Culture of Anchorage Dependent Cells, in: Develop. Biol. Standard. 42 (ed. Inter. Assoc. Biol. Standard) p. 121, Basel, S. Karger 1978
27. Obert, H. J., Joester, K. E.: Sources of Interferon for Clinical use: β-Interferon from Fibroblasts, in: Interferon 4 (eds. Finter, N. B., Oldham, R. K.) p. 89, Amsterdam—New York—Oxford, Elsevier 1985
28. Kruse, P. F., Mediema, E.: J. Cell. Biol. *27*, 273 (1965)
29. Mann, G. F.: PAHO Bulletin VI *33*, (1972)
30. Van Wezel, A.: Nature *216*, 64 (1967)
31. Katinger, D., Scheirer, W.: Mass cultivation of mammalian cells in an airlift fermentor, in: Develop. Biol. Standard. 42 (ed. Intern. Ass. Biol. Standard.) p. 111, Basel, S. Karger 1979
32. Berg, G. J.: An integrated system for large-scale cell culture, in: Develop. Biol. Standard. 60 (ed. Intern. Ass. Biol. Standard.) Basel, S. Karger 1984
33. Knazek, R. A.: Science *178*, 65 (1972)
34. Knight, E.: Proc. Natl. Acad. Sci. U.S.A. *73*, 520 (1967)
35. Berthold, W., Tan, C., Tan, Y. H.: J. Biol. Chem. *253*, 5206 (1978)
36. Heine, J. W., De Ley, M., Van Damme, J., Billiau, A., De Somer, P.: Ann. N.Y. Acad. Sci. *350*, 364 (1980)
37. Friesen, N. J., Stein, S., Evinger, M., Familletti, P. C., Moschera, J., Meienhofer, J., Shively, J., Pestka, S.: Arch. Biochem. Biophys. *206*, 432 (1981)
38. Knight, E., Fahey, O.: J. Biol. Chem. *256*, 3609 (1981)
39. Novick, D., Eshhar, Z., Gigi, O., Marks, Z., Revel, M., Rubinstein, M.: J. Gen. Virol. *64*, 905 (1983)
40. Berg, K., Ogburn, C. A., Paucker, K., Mogensen, K. E., Cantell, K.: J. Immunol. *114*, 640 (1975)
41. Hochkeppel, H. K., Menge, V., Collins, J.: Nature *291*, 500 (1981)
42. Zoon, K.: Interferon Test Procedure: Points to consider in the production and testing of interferon intended for investigational use in humans, Office of Biologics National Center for Drugs and Biologics, July 28, 1983
43. Mun, H. NG., Vilcek, J.: Adv. Prot. Chem. *26*, 173 (1972)
44. Edy, V. G., Braude, I. A., De Clercq, E., Billiau, A., De Somer, A.: J. Gen. Virol. *33*, 517 (1976)
45. Heine, K. J. W., Billiau, A.: Purification of human fibroblast interferon by adsorption to controlled-pore glass and zinc-chelate chromatography, in: Methods in Enzymology, 78 (ed. Pestka, S.) p. 448, New York, London, 1981
46. Heine, J. W., Van Damme, J., De Ley, M., Billiau, A., De Somer, P.: J. Gen. Virol. *54*, 47 (1981)
47. Edy, V. G., Billiau, A., De Somer, P.: J. Biol. Chem. *252*, 5939 (1977)
48. Havell, E. A., Vilcek, J.: Antimicrob. Agents Chemothe. *2*, 476 (1972)
49. Allison, L. M. C., Mann, G. F., Perkins, F. T., Zuckerman, A. J.: J. Biol. Standard. *9*, 185 (1981)
50. Code of Federal Regulations 21: Parts 600–789 Paragraph 630. 13 (C.2). Revised April 1, 1983
51. WHO Tech. Rep. Ser. 610, 54 (1977)
52. WHO Tech. Rep. Ser. 658, 313 (1981)

53. Morandi, M., Stanghellini, L., Valeri, A.: Problems involved in the large-scale production of biological products, such as β-interferon, using diploid fibroblast cells as substrate, in: Develòp. Biol. Standard. 60 (ed. Intern. Ass. Biol. Standard.) p. 405, Basel, S. Karger, 1984
54. Rubinstein, S., Familletti, P. C., Pestka, S.: J. Virol. *37*, 755 (1981)

Gene Synthesis

Joachim Engels[1] and Eugen Uhlmann[2]
[1] Institut f. Org. Chemie, University of Frankfurt, Niederurseler Hang, D-6000 Frankfurt 50, FRG
[2] Hoechst AG, D-6230 Frankfurt/M. 80, FRG

The scope of this review is to show the impact of chemistry on the biological sciences by presenting the basic concepts of oligonucleotide chemistry underlying gene synthesis. We demonstrate the P(III) and P(V) chemistry for chain elongation, then continue with the isolation and characterisation of synthetic oligonucleotides. The applications of oligonucleotides as tools for gene isolation, detection of genetic diseases, and site specific mutagenesis will be touched upon only in a few examples. But we review in detail the more powerful applications: the combination of synthetic oligonucleotides into double-stranded gene fragments and genes. Genes are either synthesized by the classical Khorana approach, by making use of DNA polymerase or some other specialized methods. Their design is discussed with the implications of secondary structure as well as retrosynthetic considerations. Furthermore, considerations concerning the mutagenesis of genes as well as their mode of expression including synthetic control units are outlined.

1 Introduction

In the years ahead, genes are going to be synthesized. The next steps would be to learn to manipulate the information content of genes and to learn to insert them into and delete them from the genetic systems. When, in the distant future, all this comes to pass, the temptation to change our biology will be very strong.

H. G. Khorana, 1968

We want to draw attention to the fact that recent developments in the field of chemical methodology which allows us to modulate the molecular basis of living matter is rapidly changing the face of biotechnology and biochemistry. Chemical synthesis of oligodeoxyribonucleotides, once a very specialized field for experts, has now become a central technique in biology. Through the rapid development of cloning techniques the creation of new combinations of DNA is possible be it through synthetic DNA or reorganization of existing DNA. In the last decade nucleotide chemistry has acquired a tremendous impetus. The elaboration of solid-phase techniques and the improvements in yield and speed finally resulted in fully automated oligonucleotide synthesizers. Almost quantitative yields pushed the target molecules above the hundredmers, a result one only could have dreamed of 20 years ago. These synthetic oligonucleotides in combination with DNA-processing enzymes like ligases and polymerases have opened the way for man-made genes.

2 Application of Synthetic Oligodeoxynucleotides in Biology and Medicine

According to the base pairing concept first described by Watson and Crick [1] oligonucleotides can be used as specific hybridization probes for complementary DNA or RNA. By knowing part of the amino acid sequence of the protein of interest its DNA sequence could be detected by hybridization in a colony [2] or plaque [3] assay.

2.1 Hybridization Probes

Experimental results [4–6] from single base pair mismatches show that the perfect matching sequence melts at the highest temperature, thus the type of mismatch shows individual impact on destabilization [7], a G-T mismatch and a G-A mismatch only slightly weaken the duplex. In order to guarantee a perfect match, the so-called mixed sequence hybridization probe (Fig. 1) has been introduced which was first used for β2-microglobulin [8], and has been applied many times with success. The chain length of these probes is in the range of 15–20 residues. These probes are synthesized as outlined in a mixed fashion in order to avoid multi-synthesis. The reactive nucleotides

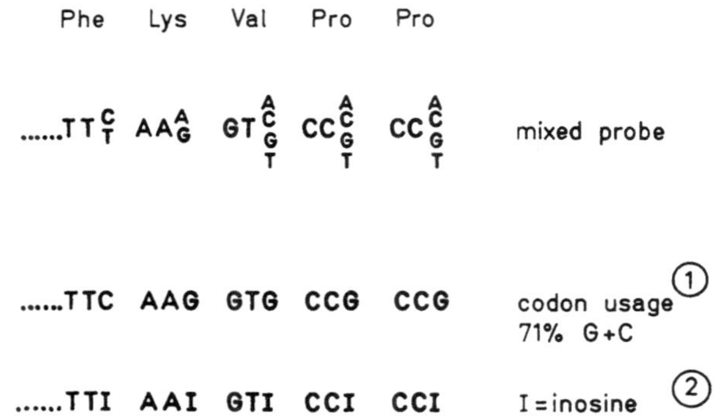

Fig. 1. Hybridization probes derived from the amino acid sequence; 1) part of a DNA probe for the 2,5-di-Keto-D-gulonic acid reductase from corynebacterium [14]; 2) hypothetical deoxyinosine containing DNA probe

are mixed for the synthesis in a 1:1 or 1:1:1 or 1:1:1:1 mixture of A, C, G, T and added to the oligonucleotide according to the sequence. This approach is based on the fact that the two purines as well as the two pyrimidines do react at essentially the same rate [9]. The limit of degeneracy is difficult to assign, — however — a 23-mer oligonucleotide mixture composed of 384 sequences has been successfully used [10]. But this does not necessarily prove that the exactly matching oligonucleotide was present and therefore was the cause of finding the wanted gene.

With the improvement of oligonucleotide synthesis, the application of long unique hybridization probes has become feasible [11]. They have been designed using the existing peptide sequence [12,13]. This approach has also been successful in the case of the vitamin C-synthesis in *Erwinia* [14] where codon usage was applied with two 41-mers. In certain organisms there is either a known statistical evaluation of codon usage [15] or as in this case a strong preference for G/C-rich codons. By this strategy the extent of mismatch has been in the order of 20%, which was enough for a specific signal in the assay. In most cases it has been advantageous to construct more than one probe for hybridization from the known peptide sequence.

Recently another type of probe places in the wobble position a deoxy-inosine [16,17] which can base pair with A, C, G and T. According to the melting temperature values of the duplexes the order of stability is: C > A > G > T. Even though this base is not a true chameleon it has been successfully applied for the α-amylase inhibitor Haim II gene with a 27-mer probe [18].

2.2 Oligonucleotides as Primers

Oligonucleotides can hybridize either with DNA or RNA and act as primers for enzymatic synthesis of complementary DNA (cDNA). This has been widely exploited by the dideoxychain termination sequencing method from Sanger [19], where a specific binding is absolutely necessary. According to the known chain terminators in the assay the sequence can be read from the radioactive bands separated on a polyacrylamide gel. Besides these applications for cloning purposes oligonucleotides have been used to prime on rare RNA's when using reverse transcriptase as an enzyme. When the mRNA is of low abundance the resulting cDNA can either be cloned after being rendered double stranded or can be used as hybridization probe for a cDNA-bank. This enrichment of rare mRNA has been used for human influenza virus RNA [20]. Alternatively to the specific priming event a so-called random priming [21] has been used. By that oligonucleotides 4 to 7 residues in length which are expected to prime at least once every 100 nucleotides generated transcripts from TMV mRNA from 500 to 2000 nucleotides [22].

2.3 Oligonucleotides for Mutagenesis

Synthetic DNA mediated mutagenesis provides one of the most powerful methods of studying proteins and DNA-protein interaction. Either single-stranded oligonucleotides in combination with phages like ∅ X 174 and M13 [23] or double-stranded oligonucleotides for cassette mutagenesis have found widespread application (see 4.5).

2.4 Probes for Genetic Diseases

Based on the ingenious blotting method developed by Southern [24], oligonucleotide probes can be used to detect single base pair changes within the human genome. For example sickle-cell anemia [25] caused by a transversion (A → T) within the β-globin gene can be analyzed with labeled (e.g. radioactivity) oligonucleotides. Under stringent hybridization conditions — washing at increasing temperatures — it can be analyzed whether one or two of a family's alleles contain the mutant gene (i.e. heterozygous or homozygous). This approach has proven to be applicable also in cases of β-thalassemia [26] and α-1-antitrypsin deficiency [27]. This approach has also been used for bacteria [28].

3 Synthesis of Oligonucleotides

3.1 Historical Background of Synthetic Oligonucleotide Synthesis

An impressive number of excellent reviews [29–38] on nucleotide chemistry has appeared in the last decade as well as several monographs [39–41], which also cover experimental techniques. Beside chemical differences in synthesizing oligonucleotides the scale also determines somewhat the approach whether the oligonucleotide synthesized is used for biological purposes or for structural use (e.g. NMR or X-ray). In this article we will confine ourselves to biological applications thus working with μg- and nmol-quantities.

After the pioneering work of Todd et al. [42] on phosphotriester synthesis, Khorana and his coworkers [43,44] were able to develop phosphodiester-chemistry to an impressive height which culminated in the tyrosine suppressor transfer RNA gene. Following

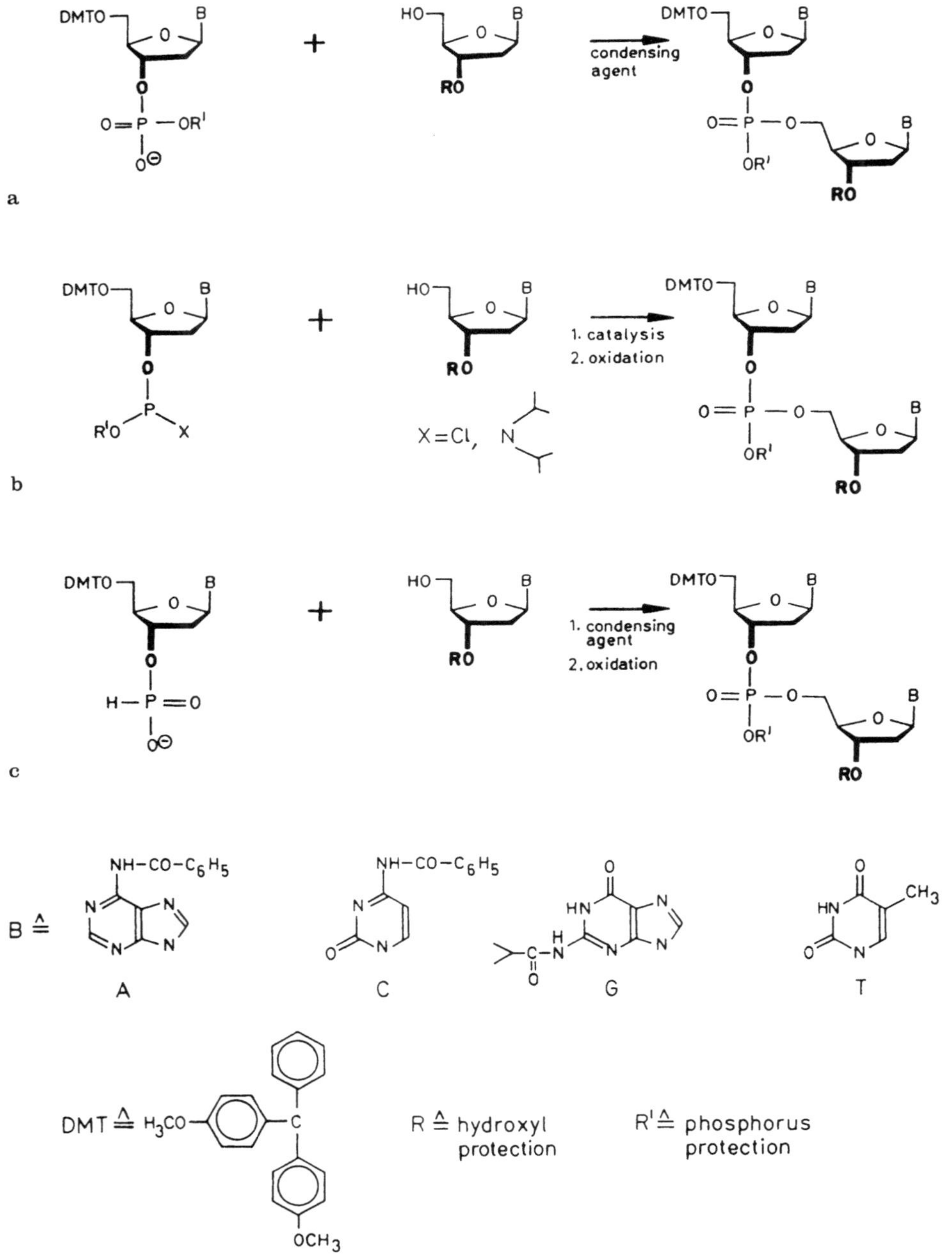

Fig. 2a–c. Three different methods for oligonucleotide synthesis; **a** triester method; **b** phosphite method; **c** H-phosphonate method

this interlude phosphotriester chemistry was revitalized by Letsinger et al. [45] and further elaborated by Eckstein et al. [46,47], Reese and Saffhill [48], Catlin and Cramer [49], Itakura et al. [50,51], Miyoshi and Itakura [52], Efimov et al. [53], Froehler and Matteucci [54] to a standard procedure.

Alternatively to this phosphorous (V) chemistry which had to be fast at room temperature, the so-called phosphite triester method, based on phosphorus (III) chemistry with higher intrinsic activity has been introduced by Letsinger et al. [55] and further elaborated on to the phosphoramidite method by Beaucage and Caruthers [56]. Recently a further alternative has been reintroduced, the so-called H-phosphonate [57–59] method which favourably compares with the above methods. This method dates back to the work of Hall et al. [60] using these phosphonates and diphenylphosphorylchloride as an activating agent to build dinucleosidephosphates. These three different methods are outlined in Fig. 2.

The following general conclusions can be made. The main difference between the triester, the H-phosphonate on one side and the phosphoramidite method on the other side is the use of a condensing agent in the first case thus rendering the system water free by excess reagent.

The final product of the synthesis before deprotection is identical — a triester — for the amidite and the triester method.

All three procedures are used for solid phase synthesis (see 3.2). Whereas methods a (Fig. 2a) and b (Fig. 2b) do share almost identical protecting groups on phosphorus the third procedure does not need further protection on phosphorus. In this respect it can be regarded as a diester analogy on phosphorus (III) chemistry.

Besides these generally accepted methods there are further modifications of the theme which have been less widely accepted. In a rapidly developing field like DNA-synthesis it certainly is difficult to aim at final solutions. Nevertheless we are trying to outline the most widely accepted techniques and choices of the reagents and conditions. The introduction of automated solid phase oligonucleotide synthesis has

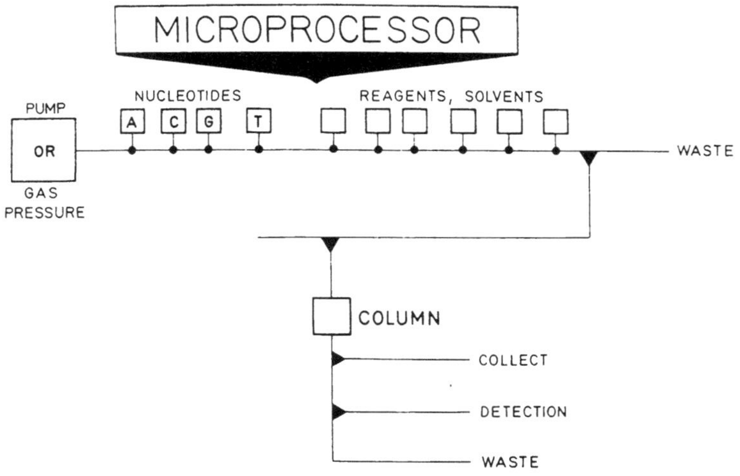

Fig. 3. Schematic representation of a microprocessor controlled oligonucleotide synthesizer with a flow through system which is operated either by a pump or pneumatically

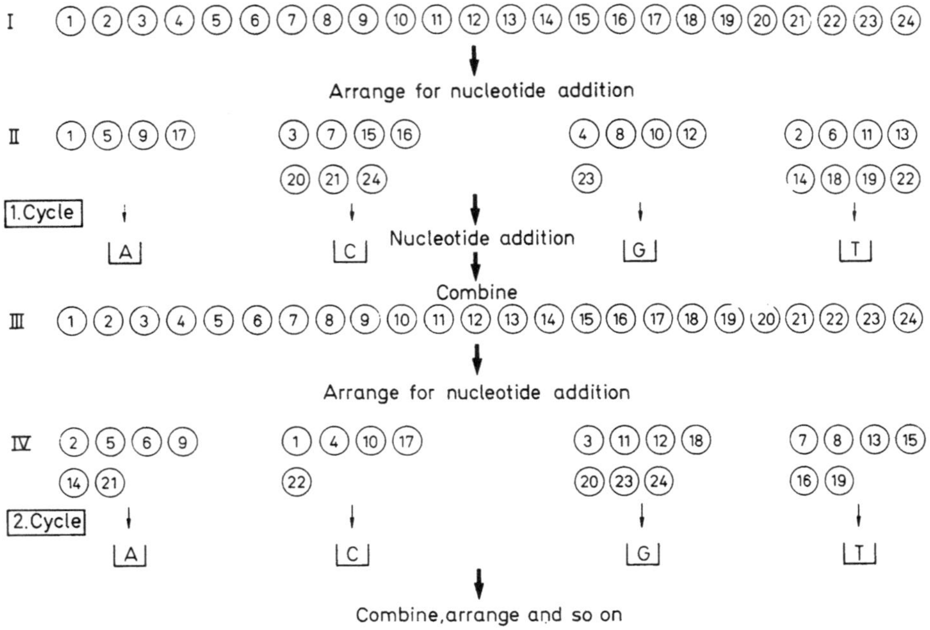

Fig. 4. Flow scheme for the simultaneous synthesis of 24 oligonucleotides by the so called "segmented technique". The circle represents numbered paper discs which are filled into the four reaction chambers A, C, G, T

streamlined the once heterogenous field of nucleotide chemistry substantially, be it by the commercial availability of starting materials and reagents, the type of synthesizer on the market or even the patent situation.

For the synthesis of oligodeoxynucleotides on polymer support two basic approaches have been followed. First the stepwise addition in a flow through type of reactor (Fig. 3) where each reaction column gives rise to an individual oligonucleotide. In the second concept, the so-called "segmented technique" [61-64] (Fig. 4), four different reaction chambers (for A, G, C, T) are used simultaneously and they contain a larger amount of polymer supports, usually paper disks. In Fig. 4 for oligonucleotide 1 in the first cycle an A will be added and in the second a C. After the addition the paper disks are combined for the washing and detritylation steps. For the first method several automats have been introduced to the market. These synthesizers use one of the above mentioned chemical methods.

3.2 Polymer Support

For the triester chemistry the Merrifield type polymer gave satisfactory results [65], especially crosslinked with 1% divinylbenzene. The inertness against organics and the mechanical stability of silica [66,67] suggested this material as the first choice in the phosphite triester approach for flow through systems. Later the use of controlled

a

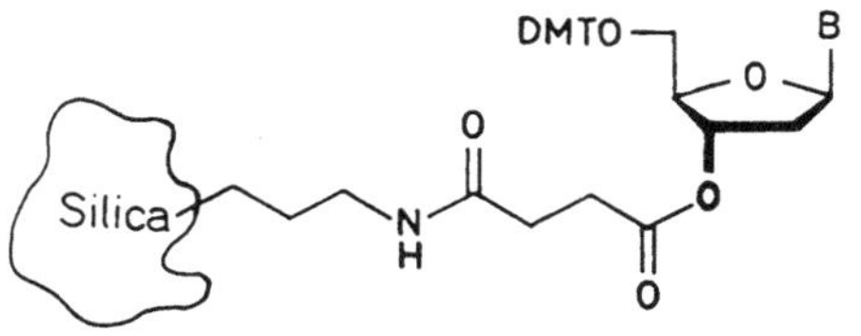

b

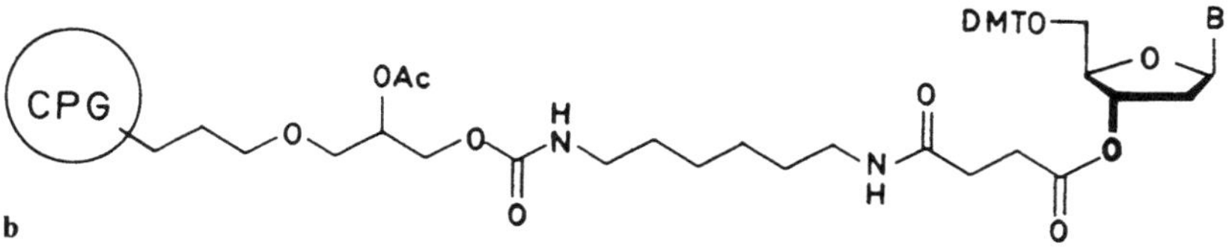

Fig. 5a and b. Supports and spacers for oligonucleotide synthesis; **a** amorphous silica, **b** controlled pore glass (CPG) beads

pore glass beads (CPG) [68, 69] proved to be even more advantageous in yield while being even superior in inertness and mechanical stability. The influence of the support on the experimentel yield can hardly be underestimated. Side reactions based on crowding and interference on the polymer severely limit the length of chain obtainable by this method. Thus the pore size of the material is of critical importance to the length of oligonucleotide. Recently with 1000 Å CPG the synthesis of oligonucleotides up to 180 nucleotides in length have been reported and up to 182 nucleotides have been prepared in the authors laboratories. The first nucleoside (Fig. 5) is hooked on the support via side chains (so-called spacers). First an aminopropyl side chain was introduced, later a longer alkyl group was favoured and introduced commercially. Besides this now favorite support a variety of other materials have been tested like cellulose [70], polyacrylamide [71] and ^R^teflon (Du Pont de Nemours) [72]. Paper as support, especially in the segmental approach has been discussed in Sect. 3.1.

3.3 Protecting Groups in Oligonucleotide Synthesis

Assembling an oligonucleotide starts with preparing the appropriately protected nucleoside derivates. Mutually compatible protections [73] have to be found not only for the heterocyclic base but also for the sugar moiety. The heights and pitfalls of nucleotide chemistry very much is determined by the appropriate choice of compatible protecting groups.

3.3.1 Protection of the Heterocyclic Bases

3.3.1.1 Protection for Adenine and Cytosine

The exocyclic amine function of adenine is less reactive than that of cytosine. The latter can be directly benzoylated in methanol under reflux with benzoic anhydride [74]. Different acyl protecting groups for the exocyclic amino group have already been prepared by Khorana et al. [75] and tested against ammonia cleavage, the reagent of choice at present.

In the adenine case the exocyclic amino function has also been protected by acylation. Here a drastic acid lability of the glycosidic bond has been observed [76] which is due to an altered protonation of the protected adenine. Cytosine as well as adenine are prepared as N-benzoylated derivatives by two procedures. Either by peracylation of the nucleosides followed by cleaving off the ester groups with ammonia [77]. Alternatively a transient protection of the nucleoside hydroxyls with trimethylsilyl groups followed by acylation has been developed [78], which is now the method of choice.

A wide variety of other protecting groups [38] for the amino function has been used, mostly borrowed from peptide synthesis. They are selectively cleaved by β-elimination, reduction and other specific reactions.

A promising alternative to circumvent the depurination problem for adenine seems to be the introduction of hindered amidine functions [79, 80] or alternatively a diacylation on N6 [81].

3.3.1.2 Protection for Guanine and Thymine

Thymine has been used unprotected in most cases. By strong phosphitylating agents [82, 83] or condensing agents [84] a reaction at the amide structure occurs. Therefore several amide protecting groups either aryloxy [85] or those cleavable by β-elimination like p-nitrophenylethyl [86], p-nitrophenylsulfonylethyl [87] have been suggested.

Guanine represents the most complicated case since the exocyclic amino group as well as the amide function is prone to side reaction. It is protected by acylation of the 2-amino group in analogy to cytosine and adenine. In the case of guanine a benzoyl group proved to be too stable to be removed by mild ammonia treatment [88]; therefore the isobutyroyl group has been introduced and is used up to now. Protection of the amide function has been intensively studied. Here again groups to be removed by β-elimination like cyanoethyl [89] and 4-nitrophenylethyl [90] as well as carbamoyl groups [91] are mostly favoured. A distinction in reactivity between the phosphorous triester- and the phosphite triester-intermediates makes this type of protection necessary in the first case, whereas optimized reaction cycles in the phosphite approach seem to circumvent unwanted side products. Further studies, analyzing more precisely the side products concerning guanine [92] will certainly clear up the picture in the future. Nowadays by optimized reaction cycles oligonucleotides well above 100 can be obtained without additional 0-6-protection, in which case the deprotection also has to be absolutely clean and efficient.

3.3.2 Protection of the Deoxyribose Moiety

The majority of synthetic strategies for oligonucleotide synthesis favors an acid labile protection of the 5′-OH group of deoxyribose. From these mostly the 4,4′-dimethoxytrityl group (DMT) has been used. It has the advantage of being introduced quite regioselectively and renders the nucleosides lipophilic enough to allow use of acetonitrile or methylene chloride as organic solvents. In addition, monitoring the release of this group as an orangered cation [498 nm] offers an easy follow up test for the efficiency of coupling reactions. Less frequently used is the pixyl group which has been introduced as an alternative for yielding crystalline building blocks [93]. The acid sensitivity of these groups is comparable, the two methoxy groups render the standard trityl protecting group ideal for nucleotide synthesis [94]. A plethora of Broenstedt and Lewis-acids [36,95] has been envisaged for cleaving this ether linkage. Dichloro- or trichloroacetic acid in methylenchloride [96,97] are the reagents of choice. Zinc bromide [98,99] though more selective regarding depurination created problems in automated syntheses.

3.3.3 Protection on Phosphorus

Regarding the strategic methodology we can divide protecting groups into two categories: permanent or persistent and temporary or intermediate. In addition we adopt Merrifield's [100] orthogonal classification: it is applied to combinations where each protecting group is removed by principles specific to each group. The intermediate 4,4'-dimethoxytrityl group and the permanent base protecting as well as phosphate protecting groups are orthogonal. On the other hand, there has been a trend towards greater simplicity with the aim of splitting off permanent blocking groups with one reagent only, preferentially ammonia.

Since several different procedures for phosphorylation or phosphitylation have been used the two approaches for P(III) and P(V) are considered separately. In the P(V) triester approach the aryloxy-function as a permanent protecting group has been used most widely, the o-chlorophenyl [101], p-chlorophenyl [102] being favored. These aryloxy-groups render the intermediate phosphoryl derivative active enough for the second phosphorylation step. By the introduction of the oximate cleavage [103] some former problems of unwanted chain cleavage in using alkaline hydrolysis conditions could be avoided. This makes the o- and p-chlorophenyl protecting groups the number one choice for the triester approach.

3.4 Introduction of Phosphorus (V) into Nucleosides

As to the method of introducing the phosphate moiety two main procedures have been advocated. They differ by using either a monofunctional or a bifunctional phosphorylating agent. Monofunctionality results after isolating and purifying in the fully protected nucleoside phosphotriester. One of the protecting groups is in the category of a permanent the other of an intermediate protecting group. Although a wide variety of different combinations has been tested, the β-halogenethyl- and the cyanoethyl-group as an intermediate protecting group has become most popular since they result in an isolable nucleotide. As to the activation of the phosphoryl group, the chloridates were checked initially giving always some unwanted side products even in pyridine. With Narang's introduction of the bis-triazolide [104] a very clean reaction appeared. In this procedure an arylphosphoric esterdichloridate reacts with triazole in the presence of triethylamine. This bifunctional phosphorylating agent as illustrated in Fig. 6 can react directly with a nucleoside. The resulting intermediate can be transformed into a nucleotide by condensation with cyanoethanol. Alternatively, it can be hydrolysed to the diester salt or reacted with a 5'-hydroxy nucleoside to form a 3'-5'-dinucleoside monophosphate. Alcohols cleavable by β-elimination [105] are added in the second step. These bis-triazolides do change the reactivity of the phosphorylating species, avoiding the formation of a symmetrical nucleoside phos-

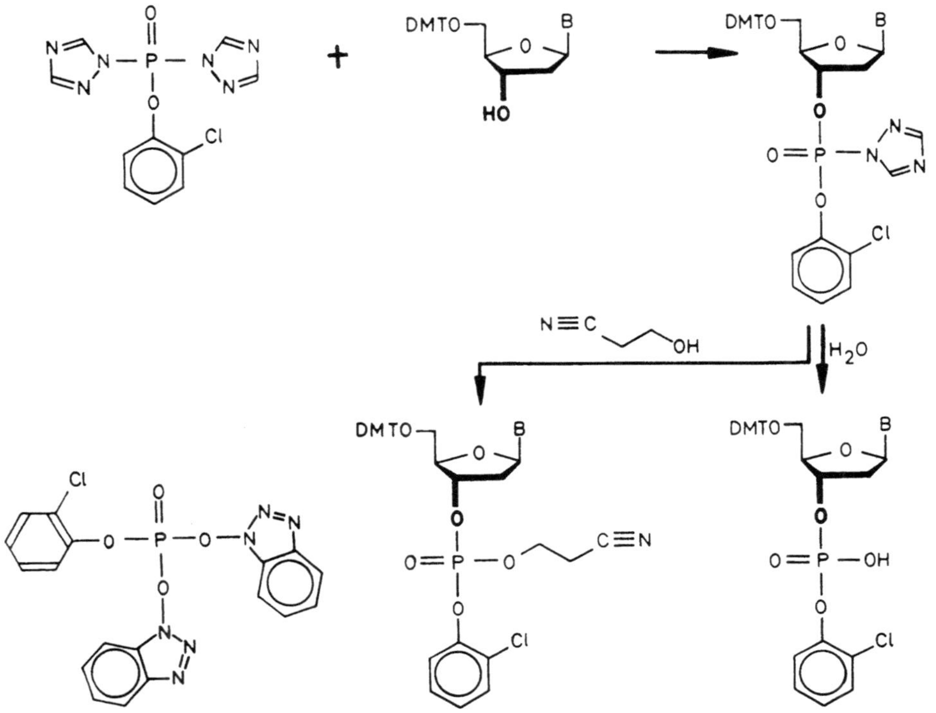

Fig. 6. Bifunctional phosphorylating agents. The bistriazolide [104] and the hydroxybenzotriazole [107] approach

phoric triester [106]. Later, van Boom introduced as a further alternative the hydroxybenzotriazole derivatives [107]. These reagents are slightly more reactive but still predominantly result in a clean phosphorylation of the nucleoside 3′-hydroxyl groups (Fig. 6) without concomitant formation of 3′-3′-dinucleoside phosphate. This procedure has been worked out to a fully automatable oligonucleotide synthesis [108].

3.5 Methods in Oligonucleotide Synthesis

3.5.1 Diester Method

By this approach, which has been elegantly demonstrated by Khorana [43, 44], a nucleoside phosphoric acid monoester with the aid of DCC reacts with a second suitably protected nucleoside to from a dinucleoside phosphoric acid diester [27, 28]. The advantage lies in the fact that no extra protection on phosphorus is needed, the disadvantage being unwanted side reactions on phosphorus.

3.5.2 Triester Method

In the classical phosphotriester synthesis (Fig. 2a) a suitably protected nucleoside phosphate was condensed with a second protected nucleotide to achieve only the desired products. The condensing agents have been the focus of much research. Sulfonic acids being the first choice had to be modified in such a way as to suppress the sulfonylation of the free hydroxyl groups. The arenesulfonyl reagents, introduced by Khorana, were modified to different azolides, the tetrazolides [109] and 3-nitro-triazolides [110], a structure which we have verified by X-ray crystallography [111]. As sterically hindered sulfonic acid derivatives either triisopropylbenzene-, mesitylene-, or chinoline-8-sulfonic acids were studied. Since the rate of the reaction very much depends on the catalyst present, most notably N-methylimidazol, 4-dimethylamino-pyridine and pyridine-N-oxide are favoured [112]. With these catalysts the reactions can be driven to completion in minutes compared to hours in earlier procedures [113]. Alternatively, these catalysts can be used in an intramolecular fashion [54].

The hydroxybenzotriazole method by van Boom does not require an extra coupling reagent neither does the phosphoramidite method. In the former N-methylimidazole is the catalyst, in the latter tetrazole is used.

3.5.3 Phosphorchloridite Method

In the phosphitylation procedure on the other hand the distinction between the two reactive halogens is much more difficult. Even at low temperature the phosphorous dichloridites do react to a symmetrical dinucleoside phosphite even in the presence of sterically demanding protecting groups. Initially by introducing the methyldi-chlorophosphites [114] a methyl protecting group was found which is cleaved by powerful nucleophiles as thiophenol and led to a synthetic strategy for oligonucleotide synthesis on polymer support [98].

3.5.4 Phosphoramidite Method

The handling of these dichlorophosphite phosphitylating agents was very tricky. By differentiating the two reactive groups on phosphorus [115] an oligonucleotide synthesis based on P(III) was finally possible as shown in Fig. 7. The permanent blocking was still done by methyl, as intermediate a secondary amine was chosen. The dimethylamino group was replaced by N,N-diisopropylamine [69] adding enough stability to the reagent for chromatographical purposes. Adding tetrazole, a mild acid, to the phosphoramidite a very clean substitution and almost quantitative phosphitylation with a second nucleoside was achieved. Recently the β-cyanoethyl [116] group has replaced the methyl-group as the most popular permanent protecting group for phosphitylation (Fig. 7).

3.5.5 H-Phosphonate Method

The H-phosphonate [57–59] approach as illustrated in Fig. 8 is undergoing at the moment the latest elaborations. It achieves differentiation of the two potentially reactive hydroxyl groups on phosphorous and arrives at stable building blocks for automated oligonucleotide synthesis. The nucleoside-H-phosphonate is activated by an acid derivative, most often pivaloylchloride in the presence of pyridine, and led to react with the second nucleoside. In automated synthesis neither an extra protecting group on phosphorus nor oxidation after each individual addition is necessary, using this approach. A key problem in this approach still remains the stability of the H-phosphonate diester towards pivaloylchloride [117]. Besides this weak point the

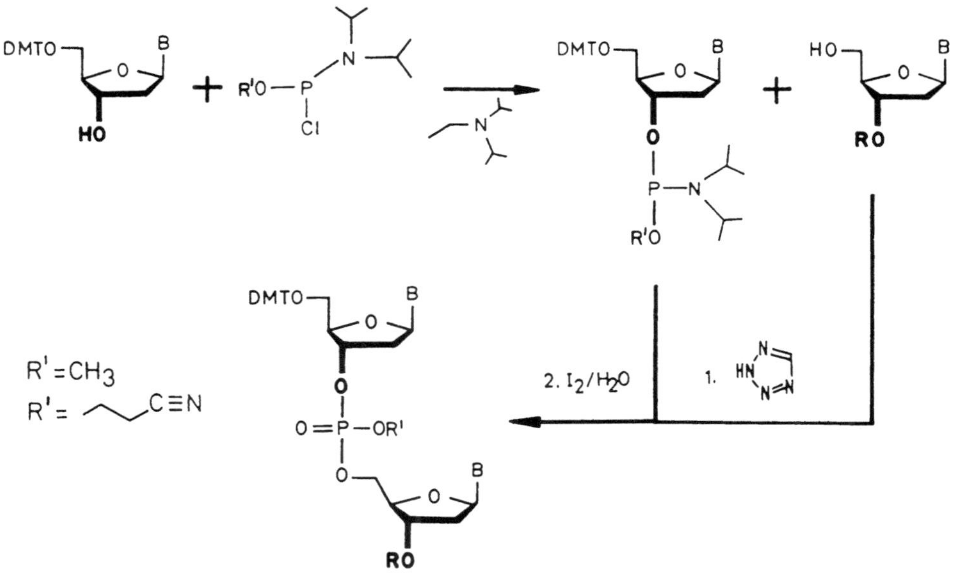

Fig. 7. Phosphoramidite based oligonucleotide synthesis. The methyl [56] and the cyanoethyl [111] variations are represented

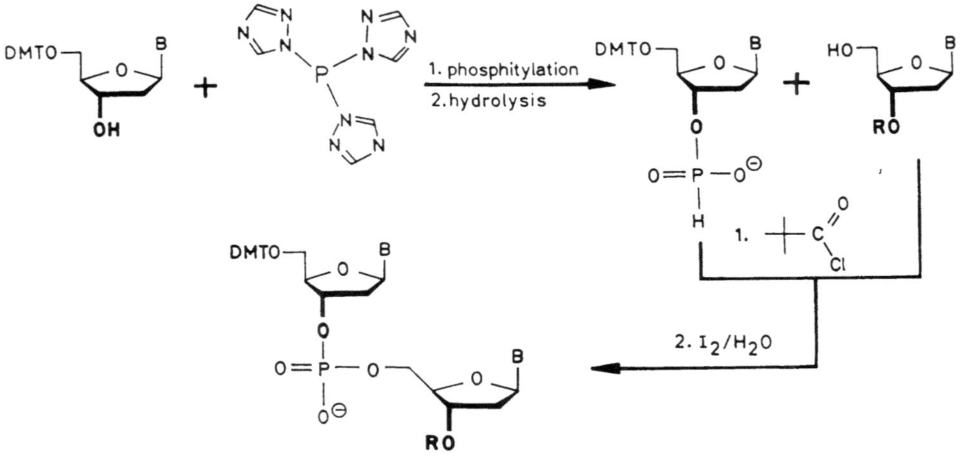

Fig. 8. Oligonucleotide synthesis based on the H-phosphonate approach

approach solves the protection group problem with a minimum of steps and easy handling.

3.5.6 Solid Phase Phosphoramidite Chemistry

In the preceding chapters we have outlined basic oligonucleotide chemistry. These different methods of generating a phosphodiester bond can either be done in solution

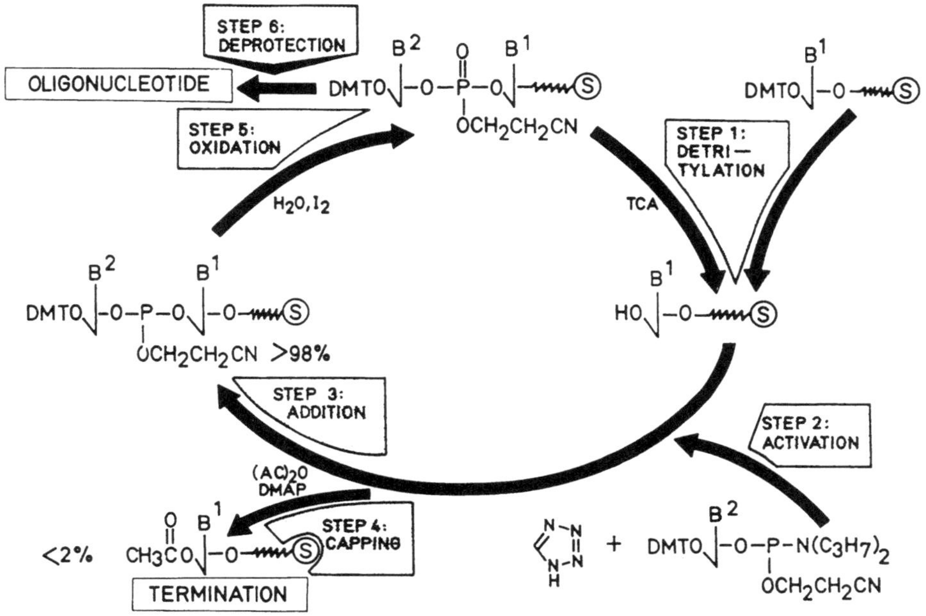

Fig. 9. Reaction cycle for a solid support oligonucleotide synthesis based on phosphoramidite chemistry

by a stepwise fashion or by using preformed blocks of dimers or trimers. Most of these methods have been performed in solution in earlier times whereas the solid phase synthesis has dominated the field. Let us consider now the arrangement of the synthesis cycle for solid phase chemistry. As illustrated in Fig. 9, the starting nucleoside is hooked on to the polymer support. With very few exceptions always the 3′-end is fixed to the polymer, leaving the 5′-position open for chain elongation. The cycle has the following features:

1. Deprotection of the 5′-position (acid).
2., 3. Activation and addition or coupling.
4. Capping,
5. Oxidation (in the case of P(III) compounds).

Between these main steps a variety of washings is done. Their main purposes are the equilibration of the medium with the solvent of the coming reaction and to get rid of all extra reagents and possible side products afterwards. It is important to guarantee a waterfree addition or coupling step, the latter being simplified by excess coupling reagent.

Step number 4, the capping, has been a matter of much discussion. Why capping when one is dealing with almost quantitative reactions? There certainly is a point in this argument. On the other hand if the addition is not quantitative, capping provides a life belt. Most researchers though tempted to eliminate this step have reintroduced it again with good results since it may reverse side reactions on the heterocyclic moiety. The oxidation in the case of the P(III) compounds has routinely been done with aqueous iodine solutions. Even though a non aqueous oxidation [118] would be ideal, none of the proposed reagents has proved to be superior to the established method so far.

3.6 5′-Phosphorylation of Synthetic Oligonucleotides

The synthesis cycle just described is repeated until the last nucleotide in the sequence has been added. The 5′-end of an oligonucleotide for DNA-probes, or linkers, or gene fragments usually exhibits a free OH-group. When ligase is used as the joining enzyme a phosphate has to be introduced into this position. This used to be done enzymatically by polynucleotide kinase. In order to simplify this step we [119, 120] and others [121–123] recently introduced several phosphitylating agents based on the phosphite approach as illustrated in Fig. 10. We were thus able to make a 5′-phosphate directly available within the synthesis cycle. After oxidation the oligonucleotide is ready for deprotection.

Of course it is possible to add other groups instead of phosphate [124]. In Sanger's dideoxy sequencing, fluorescence labels have been introduced into the oligonucleotide instead of radioactivity. In order to attach these dyes to the 5′-end of a synthetic oligonucleotide a so-called "amino link" was developed. By this an aminoethylphosphate side chain is added to the 5′-hydroxyl group of the synthetic oligonucleotide. With the help of the amino group the fluorescent dye can be attached. Furthermore, even a tritylation on the support could be accomplished [125].

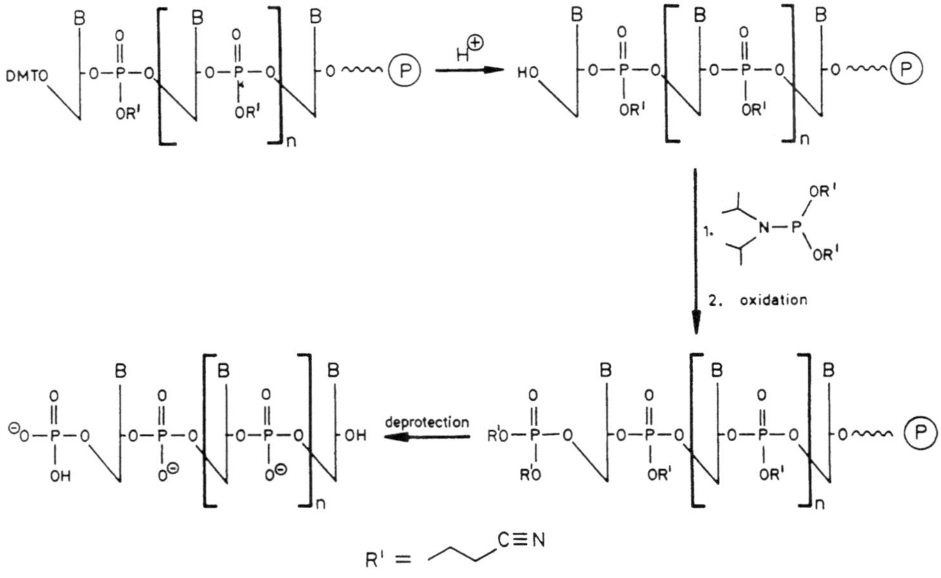

Fig. 10. Reaction scheme for the chemical 5′-phosphorylation of oligonucleotides based on phosphoramidite chemistry [119, 120]

3.7 Deprotection of Oligonucleotides

Following the synthesis cycle furnishes the fully protected oligonucleotide. As mentioned earlier, there is a trend towards simplification of the deprotection. The cleavage of all the protecting groups with concentrated ammonia either at room temperature for 48 h or at 50°–60 °C for several hours is one of the solutions suggested. In the case of the nonphosphorylated 5-position the order of deprotection is the following:
1. Cleavage of the permanent protecting groups on phosphorus.
2. Cleavage from the support.
3. Removal of the protecting groups on the heterocyclic bases.
4. Detritylation of the 5′-protecting group.

The necessity of first deprotecting the phosphate groups has been emphasized by van Boom and Reese [126], showing some base catalyzed chain cleavages at the ends, when free hydroxyl groups are available. In the case of the cyanoethyl phosphoramidites the authors did not observe a similar problem.

Whereas the first three steps usually are done in basic medium, the detritylation is an acidic step. Thus by the appropriate choice of base and protecting group they can be combined into one step as has been done also with the p-nitrophenylethylgroup and DBU [127, 128] or with the β-cyanoethylgroup and ammonia. For the H-phosphonate synthesis step 1 is obsolete and thus a simplified deprotection is possible. In this respect a trend to more base labile protecting groups on the heterocyclic moiety for the phosphite approach is observable [129, 130]. As to the phosphotriester approach the mode of deprotection with oximate, introduced by C. Reese, has already been mentioned. This procedure borrowed from the detoxification of nerve gases is accomplished by syn-2-nitrobenzaldoximate or syn-pyridine-2-carboxaldoximate [131]. This induced β-elimination requires a strong base like tetramethylguanidine. Under these conditions oligomers are cleaved from the support (succinate linkage) and some amide protecting groups are cleaved too.

For the phosphite approach, when using methyl or benzyl protection, a good nucleophile as thiophenolate [132] has to be used. Due to its unpleasant smell, its replacement by β-cyanoethyl has not left many unhappy [133].

After the preliminary steps normally the trityl or some modified trityl group remains which is cleaved by aqueous acetic acid [134] except for 5-modified ends.

3.8 Purification of Synthetic Oligonucleotides

Ideally the resulting oligonucleotide from a polymer supported synthesis could be used without prior purification. For quite a variety of purposes especially since the yields of each cycle have reached the 99% margin this has been tried with success: gene synthesis, primers for dideoxy-sequencing [135] and linkers.

3.8.1 Gel Electrophoresis

The methods of purification can be divided into two main categories, chromatography and electrophoresis. Whereas the former provides several alternatives the latter, coming from DNA technology [136], has the advantage of general applicability. Here

Table 1. Historical developments in gene synthesis

Gene for	Length of the gene (bp)	Number of oligonucl.	Method for oligonucl. synthesis	Length of synth. oligonucl.	Length overlaps	Year of appearance	Refs.
Alanine specific tRNA from yeast	77	15	Diester	5–20	4–7	1970	151)
Angiotensin II	33	7	Diester	6–11	5–6	1975	165)
Tyrosine suppressor tRNA from *E. coli*	126	26	Diester	6–13	4–7	1976	152–164)
Somatostatin	56	8	Triester	11–16	5–8	1977	166)
Insulin							
B-chain	104	18	Blocks of Triesters	10–12	4–6	1978	167)
A-chain	77	12	Blocks of Triesters	10–15	5–8		
Interferon-α	514	67	Triester/sp	14–21	7–10	1981	169)
Epidermal growth factor	170	10	Phosphoramidite/sp	11–59	7–24	1983	175)
Interferon-γ	453	66	Phosphorchloridite/sp	9–14	5–8	1984	188)
Interferon-γ	452	34	Phosphoramidite/sp	18–33	6[a]	1984	187)
Growth hormone	584	78	Phosphotriester/sp	7–26	4–10	1984	170)
Interleukin-2 receptor protein	753	38	Phosphoramidite/sp/A	27–51	6–12[a]	1986	171)
Bovine rhodopsin	1057	72	Phosphoramidite/sp/A	15–40	4–5[a]	1986	141)

sp = Solid phase synthesis;
A = Automated DNA synthesizer;
[a] Preformed duplex DNA

a polyacrylamide gel (PAGE [137,138]) consisting of 6–20% polyacrylamide and 1–2% methylenebis-acrylamide in the presence of 7 M urea is used routinely. The percentage of the PAA gel depends on the length of the oligomer to guarantee adequate mobility and it correlates linearly with the chain length in the obtainable size range [139,140]. The expected oligonucleotide is visualized by UV-shadowing with a fluorescent tlc sheet at 254 nm or stained with silver or a fluorescent dye like ethidium bromide. The expected nucleotide is usually the strongest and least moving band on the gel. It is cut out and extracted with water and buffer [136]. The solution containing the oligonucleotide is desalted either by gel chromatography, e.g. ^RSephadex G50 (Aktiebolaget Pharmacia) or on RP-columns or cartouches [141]. The solutions are lyophylized and the resulting oligonucleotide is kept at —20 °C for storage.

3.8.2 Chromatography (HPLC)

Several modes of separation are possible, according to size, charge or hydrophobicity [142]. All three methods have been used and here only the HPLC versions [143–146], which are ideal for small quantities are discussed. The ease of handling, the life span of the columns, and the prize have made the RP columns the most widely used. Either C_{18} or silica gel give good separations, especially when the DMT group on the 5′-end is still retained [147]. In order to avoid detritylation after the ammonia treatment the oligonucleotide mixture is directly applied or else the raw oligonucleotide is kept in a buffered solution. Due to its hydrophobic character the desired oligonucleotide is retained and separates well from the failure sequences up to 40-mers [148]. In addition modified trityl groups [149] with aliphatic side chains are used to further retard the compounds. Detritylated oligomers are also separated on C_{18} or C_8 columns. With longer oligonucleotides >50 nt certain problems are encountered, probably due to secondary structure.

Ion exchange chromatography, usually on strong anion exchangers, is most often used. Here the charge and the retention correlate directly, the longest oligonucleotide elutes last. Using potassium phosphate buffers (KH_2PO_4) formamide [150] has been added to prevent hydrogen bonding or acetonitrile in the case of triethylammonium phosphate.

4 Construction of Synthetic Genes

4.1 Historical Background of Gene Synthesis

In the sixties and early seventies, Khorana and his coworkers succeeded in pioneering the complete synthesis of genes whereby genes for alanine specific transfer RNA from yeast [151] and for tyrosine suppressor transfer RNA from *E. coli* [152–164] were constructed. Thereby, the basic techniques of gene synthesis were developed:

1. Chemical synthesis of oligodeoxynucleotides and appropriate purification methods;
2. the joining of single-stranded DNA fragments by virtue of the enzyme polynucleotide ligase to form DNA duplex fragments;

3. further enzymatic ligation of relatively small duplex molecules bearing protruding single stranded ends which finally yield the entire gene.

In fact, the fundamental strategies and methods employed in this early work have for a long time set the standards in gene synthesis and have been successfully used for the construction of numerous genes (Table 1). In contrast, the chemistry for the preparation of the oligodeoxynucleotides has been dramatically improved since then.

Despite the enormous efforts and achievements in the preparation of the tRNA encoding genes the major breakthrough came only with the successful construction and expression of functional genes for valuable peptides. Already in 1975, the first total synthesis of a structural gene encoding a peptide, namely the octapeptide angiotensin II [165] was achieved. Unfortunately, no expression of this gene has been reported. Without doubt, the most exciting contribution of oligonucleotide chemistry to recombinant DNA technology was the total synthesis of genes for the peptide hormone somatostatin [166] and for the insulin A and B chains [167], because these genes after insertion into an appropriate expression vector and transformation of *E. coli* could be successfully expressed to yield the corresponding DNA programmed peptides [166, 168].

In 1981, another landmark was the total synthesis of a structural gene for human leukocyte interferon (IFN-α), which consists of 166 amino acids [169]. This was the first time a synthetic gene was designed on the basis of the nucleotide sequence rather than from peptide sequence. The synthesis of the 514 bp DNA fragment encoding IFN-α was only made practicable by the development and application of the solid phase method for rapid preparation of the oligonucleotides. Synthesis in solution of a gene of such a size would not have been a realistic task. The introduction of sequential cloning techniques (see 4.2.1.4) rendered the construction of even longer genes possible. In 1984, a 584 bp gene encoding human growth hormone (192 amino acids) was constructed from 78 synthetic oligonucleotides [170]. The first example of a synthetic gene for a receptor protein, consisting of 753 bp, was build up from merely 38 oligonucleotides [171]. The longest synthetic gene known to date consists of 1057 bp and is coding for the 384 amino acid bovine rhodopsin protein [141]. The rhodopsin gene was constructed from 72 synthetic oligonucleotides and was designed to allow systematic structure function studies of the protein. In the latter two cases automated DNA synthesizers were used for the preparation of the oligonucleotides. By means of DNA synthesizers, which had a major impact on gene synthesis, successful preparation of even larger genes will be in reach with a justifiable synthetic expenditure.

So far about 100 synthetic genes were published (patents, not cited, and papers) involving hormones [165–167, 170, 172–185], interferons and lymphokines [169, 186–196] or other immunologically important peptides [171, 197, 198], growth factors [199–205], enzymes [206–210], neuropeptides [211–215], calcium binding proteins [216, 217], gastrointestinal peptides [184, 218], proteinase inhibitors [219–221] and other products [141, 151, 164, 222–225].

4.2 Construction of Double-Stranded DNA

The genetic information is deposited in the coding strand of the DNA and indirectly in the noncoding strand. Bihelical structure of DNA serves the purpose to stabilize the

DNA and is a prerequisite for replication. The natural ability of polynucleotides to align by base pairing can be used to build up double-stranded DNA from appropriate chemically synthesized oligonucleotides.

4.2.1 Khorana Strategy

4.2.1.1 Gene for a Tyrosine Suppressor Transfer RNA

In 1976, Khorana and his coworkers reported in twelve successive publications [152-163] on the total synthesis of the structural gene for the precursor of a tyrosine suppressor transfer RNA from *E. coli*. The nucleotide sequence of the synthetic gene was derived from the published structure of the 126 ribonucleotide tRNA precursor [226]. A plan for the 126 bp synthetic gene was presented [152] involving the following steps: 1) chemical synthesis of 26 oligodeoxynucleotides, 2) the joining of several 5′-kinased oligodeoxynucleotides by DNA-ligase to form a total of four DNA duplexes, 3) further ligation of the four duplexes by virtue of their 3′- or 5′-protruding single-stranded ends to give the total gene.

Five papers [153-157] comprise the chemical synthesis of the 26 oligonucleotides, which was done by the phosphodiester method and was the most time consuming part of the work. Fig. 11 shows the nucleotide sequence of the designed gene, its subdivision into four duplex DNA fragments (I–IV), which in turn are divided into 26 single-stranded fragments (oligonucleotides). Oligonucleotides 1 and 26 are not needed for the sticky ended form of the structural gene.

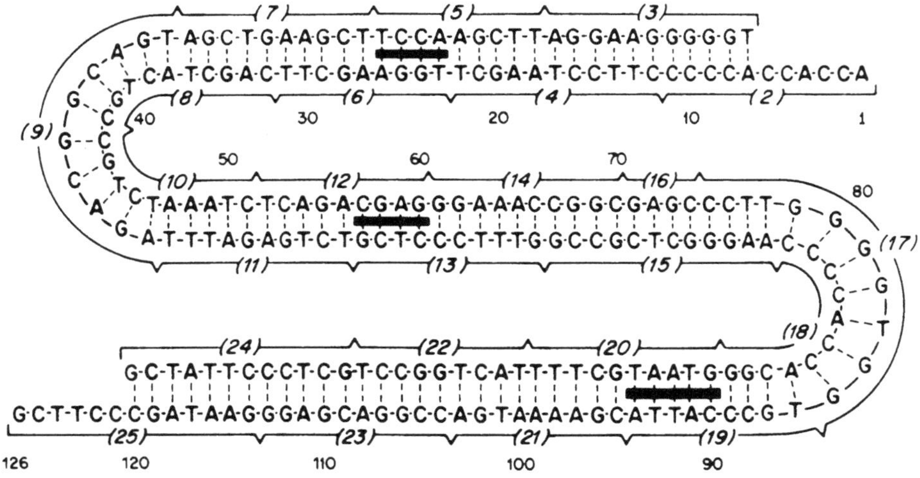

Fig. 11. The totally synthetic DNA duplex corresponding to the entire sequence (126 nucleotides) of the precursor for an *Escherichia coli* tyrosine suppressor tRNA. The numbers and distances between the carets show the oligonucleotides which were synthesized chemically. The carets indicate the sites where joining was accomplished by the use of polynucleotide ligase. The three thick strips between the duplex indicate the sites where **a** duplex [I] and duplex [II] (between nucleotides 23 and 26), **b** duplex [III] and duplex [IV] (between nucleotides 90 and 94) and **c** duplex [I + II] and duplex [III + IV] (between nucleotides 57 and 60) were joined (with permission from J. Biol. Chem. [162])

The basic requirement in dividing the duplex DNA in a retrosynthetic way was to provide for overlaps of 4 to 6 base pairs between the hybridising oligonucleotides. Of course, there are many ways to do so. Therefore, the DNA was divided so as to achieve a maximum economy in chemical synthesis, e.g. multiple use of oligonucleotide building blocks. Additionally, selfcomplementary sequences were avoided. During the enzymatic joining of the synthetic oligonucleotides [158-161], it was realized that the yields in joining reactions are unpredictable, varying from 0 to 90%. In some cases the inhibition of joining could be explained by formation of "selfstructures" or "hairpins" (see also 5.1.3). Using all possible combinations of three or more oligonucleotides at a time in the ligase reaction to get duplex II it has been found that yields were optimal when six or more oligonucleotides were ligated in one pot ("more is better") [164]. The internal oligonucleotides were phosphorylated enzymatically at their 5'-hydroxy groups prior to use in the joining reaction whereas the terminal oligonucleotides were not. After preparation and isolation of the duplex DNA, it was characterized by enzymatic degradation. Three methods were used [158]: 1) determination of resistance to bacterial alkaline phosphomonoesterase, 2) degradation to 3'-nucleotides (micrococcal nuclease/spleen phosphodiesterase) and 3) degradation to 5'-nucleotides (DNase/snake venom phosphodiesterase). The DNA duplexes I–IV were then combined in two steps to form the total gene [162]. Enzymatic joining of the duplex molecules was more rapid and efficient compared to the ligation of the single-stranded oligonucleotides. Later on, the biological activity and transcription in vitro of the cloned gene could be demonstrated [227].

4.2.1.2 Genes for Human Insulin A and B Chain

Only one year after the publication of a chemically synthesized gene for somatostatin [166] the same group around K. Itakura reported the chemical synthesis of genes for the human insulin B chain (104 bp) and A chain (77 bp) [167].

Synthesis of the 29 required oligonucleotides was performed by a block phosphotriester method using trinucleotides as building blocks. The 45 different nucleotide trimers necessary for the construction of the two genes were synthesized in three months and the synthesis of the 29 oligonucleotides, ranging in size from 10 to 15 nucleotides, took another month. The nucleotide sequence was derived from the known amino acid sequence of human insulin. The codons within the gene were chosen considering following restrictions. First, amino acid codons thought to be favored for expression in *E. coli* were used where appropriate [229]. Second, since the genes were constructed according to the Khorana strategy by ligating overlapping fragments, the oligonucleotides were designed not to undergo undesirable inter- and intramolecular pairing. Third, GC-rich followed by AT-rich sequences were avoided to eliminate potential transcription termination [230].

The synthetic genes were inserted separately into plasmids by means of appropriate single-stranded termini corresponding to the restriction enzyme sites EcoRI and BamHI. A HindIII site in the middle of the B chain gene allowed amplification of each half of the B chain gene before assembling the total B chain gene. In order to facilitate the separate expression of each of the two chains, the genes were supplied with a Met start codon on the 5'-end and two translation termination codons at the 3'-end. After successful expression of the individual peptide chains in *E. coli*, biologically active insulin was generated by disulfide-bond formation between B- and A-chain peptides in vitro [168].

4.2.1.3 Gene for Human Interferon-α_1

The total synthesis of a 514 bp gene coding for human Interferon-α_1 (166 aa residues) was another breakthrough in gene synthesis in that the state-of-the-art techniques were consequently exploited to synthesize a number of 67 oligonucleotides, ranging in size from 10 to 21 residues [169]. This has become realizable only by the introduction of the solid phase phosphotriester approach in combination with efficient purification techniques as HPLC using ion exchange or C_{18} reverse phase columns. The total gene was divided into 67 oligonucleotides so as to give overlaps of at least 7 bp on each side of all ligation points (Fig. 12). To allow the enzymatic ligation of the oligonucleotides, they were first phosphorylated by polynucleotide kinase and γ^{32}-ATP, except for the terminal oligonucleotides number 1 and 67. These oligonucleotides were grouped to give after initial ligation a set of 11 duplex DNA fragments (A–K). As an example for the first ligation step the construction of DNA duplex A from five overlapping oligonucleotides is outlined in Fig. 12a. For the next step only head-to-tail joining of I and J was designed to proceed via 5′-protruding ends, all other duplex molecules were ligated by means of 3′-protruding ends. The 11 duplex fragments were then ligated to yield the four gene fragments I–IV which were finally combined to the total interferon gene. The intermediate duplex fragments were all purified by polyacrylamide gel electrophoreses in the absence of urea using appropriately sized markers.

The interferon gene was designed from the published DNA-sequence [231] by scanning with a computer program. In order to eliminate complementary and repeated sequences within the gene a total of 71 nucleotides have been changed with respect to the native sequence without changing the encoded amino acid sequence but making the ligation of the synthetic oligonucleotides safer and more straightforward. These changes were performed considering the preferred codon usage in *E. coli*. Other features of the gene are a ATG start codon preceeding the coding region as well as a TAA stop codon at the end of the coding region. To facilitate the orientated integration of the gene into an appropriate plasmid the sticky ends for the restriction endonucleases BamHI and SalI were added to the 5′- and 3′-end of the gene, respectively.

After cloning of the gene four selected clones were subjected to analysis by restriction enzyme digestion. All four clones showed the expected pattern on PAGE. One clone was then completely sequenced and shown to be in total agreement with the initially designed sequence. In a succeeding paper the expression of the chemically synthesized human interferon-α_1 gene has been described [232].

4.2.1.4 Sequential Cloning Approach

From the early work of Khorana we have learned that a one-pot ligation reaction of six or more oligonucleotides, preferentially an even number, gives a satisfactory result. In the meantime, the assembly of a 280 bp structural gene from 20 oligonucleotides in a single ligation reaction has been reported [199]. The question of how many more oligonucleotides can successfully be ligated in a single pool stays open.

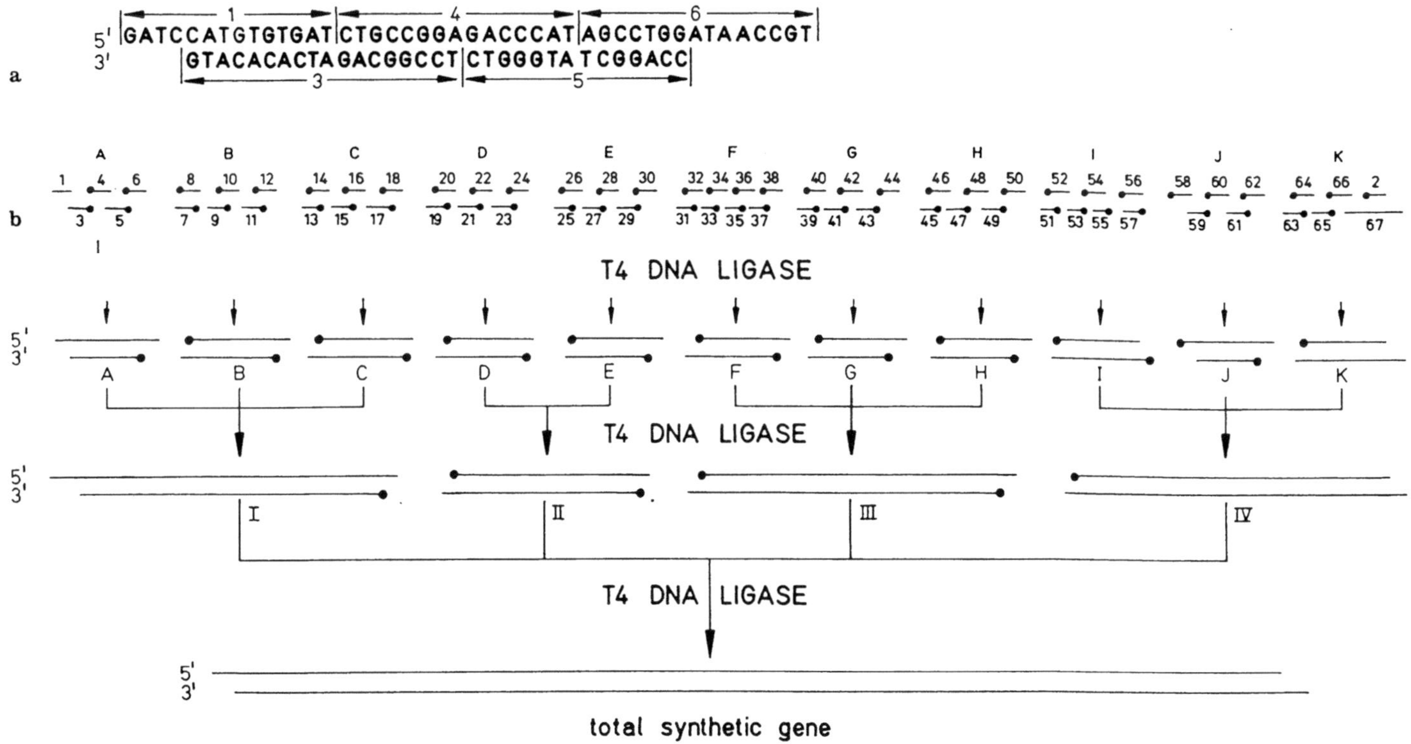

Fig. 12a and b. Ligation strategy for the INF-α_1 gene [169]; **a** example for the construction a DNA duplex (A); **b** ligation scheme of duplex DNA fragments A–K

It surely depends on the design of the specific nucleotide sequence and the quality of the oligonucleotides used.

Assuming a yield of 90% per ligation point brings the total yield in a 20 piece ligation reaction to 15% theoretical yield. In reality, some wrong annealing as well as slight impurities of oligonucleotides will decrease the yield even more. What is even more serious, at some point the mutation frequency in the synthetic gene will rise to a value that no longer or only with enormous screening effort gives a clone with the correct sequence. Of course, we can reduce the number of ligation points for a given DNA-fragment by the use of longer oligonucleotides, but with the growing length of an oligonucleotide the quality usually suffers. Therefore, it is wise to subdivide large

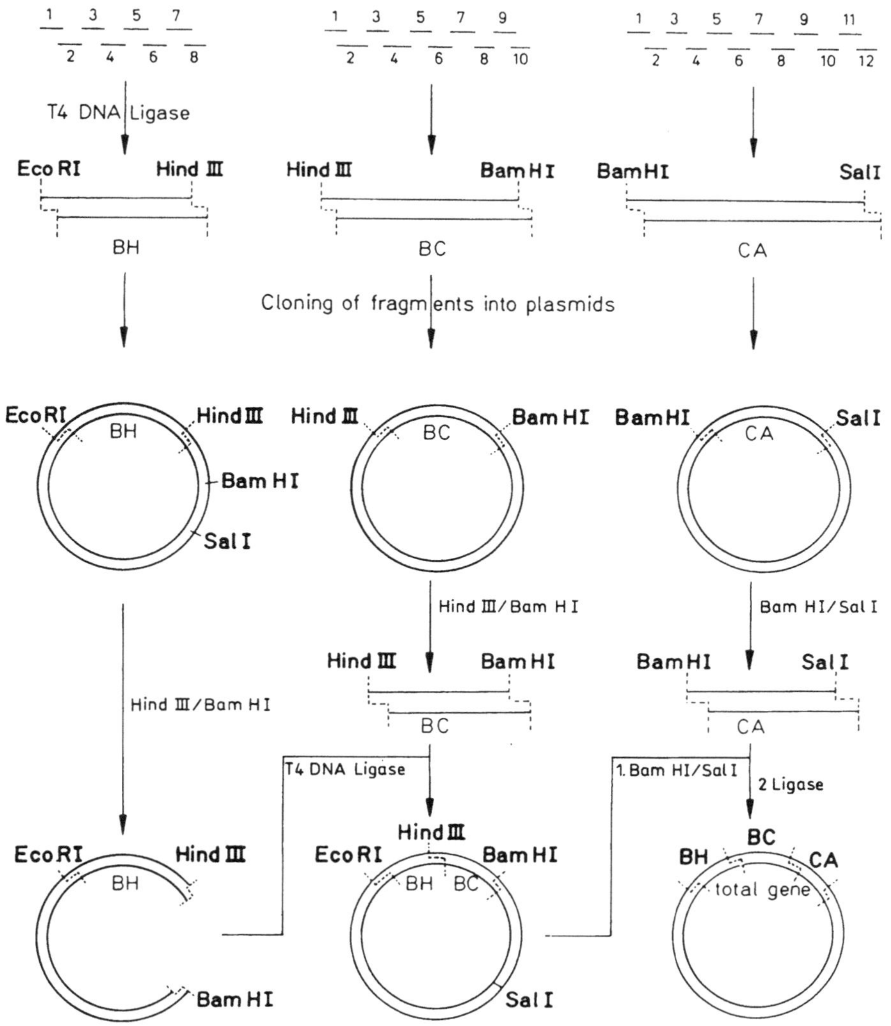

Fig. 13. Sequential cloning approach. Construction of a human proinsulin analogue gene [172]

genes into fragments which can be cloned separately in a sequential manner. An example is the synthesis of a gene for a "mini-C" analog of human proinsulin [172].

Fig. 13 shows the designed gene which could be divided into three pieces according to the two unique internal restriction enzyme sites (HindIII and BamHI). The three fragments (46, 59 and 80 bp) were constructed by ligation of 8, 10 or 12 oligonucleotides, respectively. The average length of the oligonucleotides has been 12 nucleotides, the size of the total gene was 185 bp. After separate cloning of the three subfragments BH, BC and CA they were combined to the total gene. Using the same strategy Alton et al. [186] reported the total synthesis of different human IFN-α and IFN-γ genes. The gene for IFN-α (450 bp) was divided in three parts, each approximately one third of the entire coding region and each subunit was obtained from a single one-pot ligation reaction. Similarly, genes for IFN-γ (452 bp) [187, 233] were constructed from three subfragments, the gene for human growth hormone (584 bp) [170] from 2 subclones and the gene for bovine rhodopsin (1057 bp) [141] from three subfragments. Basically, this method of sequential cloning should allow the synthesis of genes of any sizes. If there are no appropriate restriction enzyme sites within the gene, then temporary restriction enzyme sites (see 5.1.2) can be used. Furthermore, this approach is useful for combination of synthetic gene fragments with gene fragments of natural source [234].

4.2.2 Fill-in Strategy

As the progress in chemical synthesis of oligonucleotides made it possible to synthesize fragments larger than 40 nucleotides in length an alternative method to the Khorana ligation strategy for the construction of bihelical DNA fragments was published [235].

The basic principle of this so called "fill-in" method, proposed very early by Khorana [151, 236], is the synthesis of relatively large oligonucleotides of more than 40 nt units in length, which at their 3'-end share a short stretch of perfectly base pairing sequences (Fig. 14). After correct annealing of the two oligonucleotides one can make use of the DNA polymerase-mediated repair system. In the presence of the four deoxynucleotide-5'-triphosphates (dNTP) and DNA polymerase I (Klenow fragment) a full length duplex DNA fragment can be synthesized.

Using two sets of two oligonucleotides (43-/39-mer and 39-/42-mer) which were properly annealed by means of 10 and 9 base pairs, respectively, and subsequently filled in with dNTP's and Klenow enzyme, two blunt ended DNA fragments were generated. After digestion with EcoRI and PstI to create sticky ends for cloning and blunt end ligation of the two duplex DNA fragments Rossi et al. [235] succeeded in the

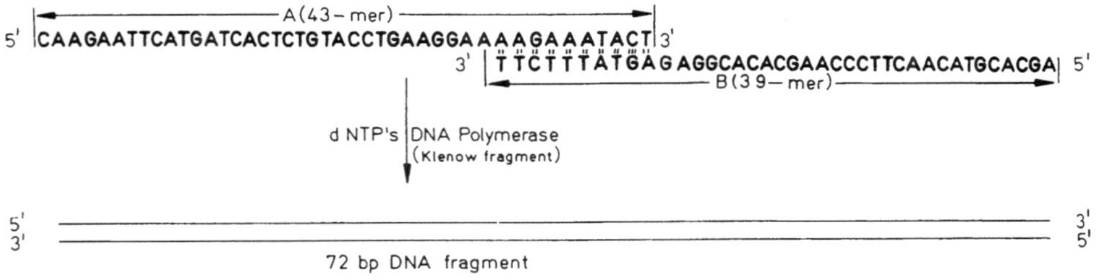

Fig. 14. "fill-in" strategy [235] using DNA polymerase (Klenow fragment)

synthesis of a 144 bp DNA fragment, whereby approximately 40% of chemical synthesis work was saved compared to the method pioneered by Khorana.

In the same year (1982), a paper fo Scarpulla et al. [237] appeared, where a similar approach was used. But it was different from the former approach in that the long oligonucleotides were preformed from short, partially overlapping oligonucleotides and AMV reverse transcriptase was used for fill in. After sequential blunt end ligation of linker molecules which could be recleaved by the corresponding enzymes to generate sticky ends, the entire assembly was cloned into pBR322. So far, only a few more papers [219, 220, 238, 239] came into our hand using this fill-in strategy, although the reduction in chemical synthesis work is substantial. From our own work [305] in the synthesis of a 100 bp DNA-fragment encoding a part of the IFN-γ we realized that the following potential problems might prevent this approach to become as standard a lab procedure as the Khorana methodology: 1) The fill-in of the last nucleotide to get the blunt end duplex proceeds only very slowly and often this nucleotide is missing. This problem can be overcome by synthesizing two to three nucleotides more than necessary and recleaving with a restriction enzyme to generate the proper ends. 2) In some instances, cleavage of the initially generated blunt ends proceeded only slowly, even when 2 or 3 additional nucleotides were present. However, it was reported that cleavage of an EcoRI linker could be speeded up by almost an order of magnitude, if the 5'-ends of the oligonucleotides were phosphorylated [241]. 3) Formation of undesired secondary structures is the most serious problem of this approach, especially, when long synthetic oligonucleotides are used (over 100 nt).

4.2.3 Direct Cloning of Long Synthetic Oligonucleotides

Nowadays, the chemical synthesis of oligonucleotides up to a length of 180 nucleotides [242] is feasible. That means, only by synthesis of two oligonucleotides, the coding strand and the noncoding strand, and subsequent hybridisation of these single stranded fragments, genes or gene fragments encoding up to 60 amino acids are accessible. Recently, the synthesis and cloning of a gene for a calcitonin derivative by this approach has been reported [243]. Using this method the work that has to be done manually is substantially reduced, because there is no need for enzymatic ligation or fill-in reaction, chemical synthesis work is done by an automated DNA synthesizer and purification of products is reduced to merely two oligonucleotides. In combination with the sequential cloning of gene subfragments this methodology is also applicable to larger genes.

4.3 Single-Stranded DNA Designed for Direct Cloning

4.3.1 Direct Ligation of Single-Stranded DNA into Appropriately Cleaved Double-stranded Vectors

Basically, the genetic information is embodied in the coding strand of a gene, whereas the second strand mainly provides for security and replication. One therefore can eliminate a great deal of synthetic work by inserting a single-stranded DNA into an appropriate vector. Subsequently, bacteria are transformed by the resulting gapped vector. This could be achieved in the synthesis of a gene fragment coding for a segment

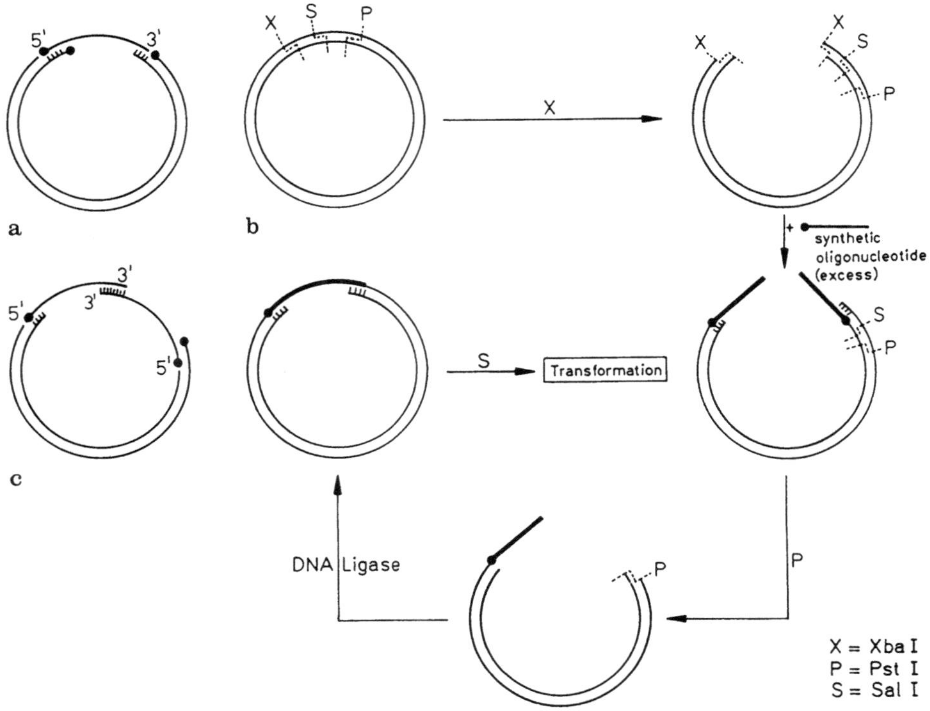

Fig. 15a–c. Single-stranded DNA cloning; **a** Direct ligation of an oligonucleotide into a plasmid with complementary 5′- and 3′-overhanging ends [244]; **b** Sequential cloning of single-stranded DNA [246]; **c** cloning of two oligonucleotides with 3′-complementary ends into a cleaved plasmid with 5′-protruding ends [244]

(aa 198–231) of bacteriorhodopsin [244]. The method makes use of the fact that single-stranded oligonucleotides with appropriate 5′- and 3′-ends can be ligated directly into double-stranded plasmids with complementary 5′- and 3′-overhanging ends in a band-aid fashion (Fig. 15a). So far, only relatively short oligonucleotides have been cloned by this method and a poor efficiency has been reported [245]. The efficiency could be increased by a fill-in reaction in vitro or by a sequential ligation process (Fig. 15b) [246]. The strategy is to cleave the plasmid first with the restriction enzyme that generates 5′-protruding ends and to force the first ligation step by a large excess of single-stranded oligonucleotide. PstI digestion removes one of the added oligonucleotides and provides a 3′-protruding end for subsequent circularization of the plasmid. A SalI digestion step before transformation cleaves and deactivates unwanted plasmids. In the latter method the cloning efficiency has been reported to be greater than 90% (oligonucleotides up to 28 nt). For gene synthesis, indeed, this approach will only be attractive if the method works for long synthetic oligonucleotides. The limitation to use a restriction enzyme pair which leaves a 5′- and a 3′-protruding end can be circumvented [244] by using two oligonucleotides which are annealed by their complementary 3′-ends (see 4.2.2). This assembly can be ligated into a vector with 5′-protruding ends at each side (Fig. 15c).

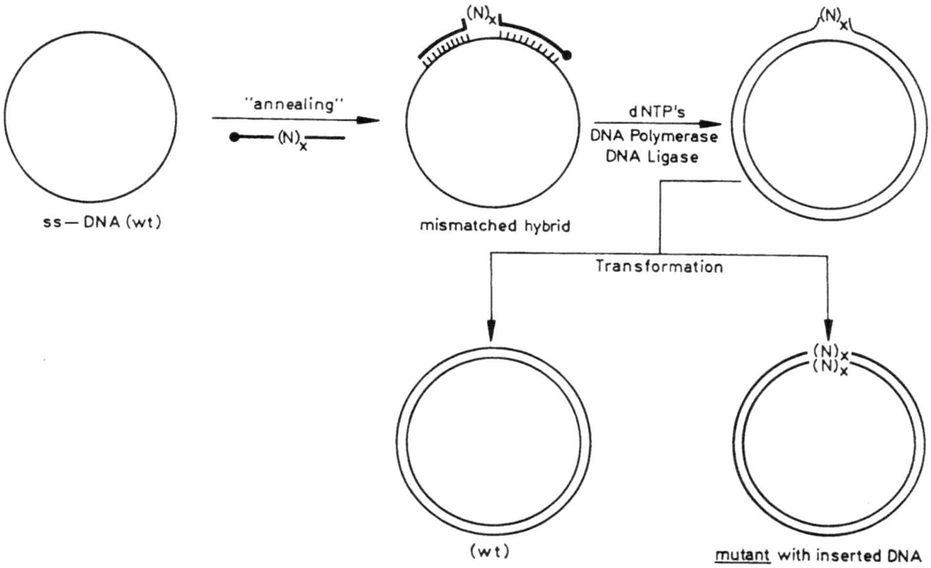

Fig. 16. M13 insertion approach. A synthetic oligonucleotide which is complementary to the ss phage DNA, except for the region $(N)_x$ to be inserted, is annealed to the ss-DNA to form a mismatched hybrid. Enzymatic fill-in and transformation yields the expected mutant plasmid

4.3.2 M13 Insertion Approach

Within the last few years, oligonucleotide-directed in vitro mutagenesis [247] has become a standard method for specific alteration of DNA sequences (see also 4.5.1). The basic principle is to use an oligonucleotide which is complementary to a single-stranded M13 template, except for a single mismatch which generates the mutation. After annealing of the oligonucleotide to the template the second strand is filled in by DNA-polymerase (Klenow fragment) and circularized by ligase. To facilitate the insertion of a DNA sequence, a loop carrying the desired sequence is in place of the mismatch (Fig. 16). Discrimination between wt plasmid and mutant plasmid having the desired DNA inserted is made by the techniques discussed above (see 2.1 and 2.4).

Now a novel strategy for total synthesis of genes is under investigation [248] which combines the improved "gapped duplex" [249] mutagenesis method with the idea outlined above. Thus, a gene of 770 bp in length is assembled by repetitive construction of insertions using the gapped duplex method and merely 11 oligonucleotides ranging in size from 65 to 103 nucleotide units. In principle, no restriction enzyme sites are necessary for cloning of the internal sequences, since the part of the oligonucleotide which is complementary to the template can be chosen more or less arbitrarily, only fulfilling certain secondary structure restrictions.

4.3.3 Ligation of Oligodeoxynucleotides by T4 RNA Ligase

The two methods of single-stranded DNA cloning as outlined in the two preceding sections are most attractive if very long oligonucleotides, e.g. one hundred or more nucleotides in length, are available. Therefore, a procedure for the combination of oligonucleotides to yield larger single-stranded oligonucleotides is requested. To date, there exists no efficient method for chemical ligation of unprotected oligonucleotides. DNA ligase is unable to catalyze this reaction, but T4 RNA ligase has been reported to accept single-stranded DNA molecules as substrates [250]. Although the efficiency of joining DNA molecules is much less than of joining RNA molecules, the reaction proceeds at optimized conditions [251, 252]. In order to prevent multiple coupling of the oligonucleotides one of them is used as 3',5'-dihydroxy component (acceptor oligonucleotide) and the other one bearing a 5'-phosphate group (donor oligonucleotide) must be 3'-hydroxy protected. Protection is usually achieved by a phosphate group which is removable by the action of alkaline phosphatase. By using polyethylenglycol and hexammine cobalt chloride as additives in the T4 RNA ligase reaction, reaction times were dramatically reduced and yields of 40 to 60% have been reported for the ligation of single-stranded DNA fragments of up to 40 bases in length [253].

An alternative approach to improve the ligation of single stranded oligonucleotides is to use those terminating in a 3'-ribonucleoside [252, 254, 255]. There is no need for 3'-protection of the oligonucleotides, provided that the joining is done starting from the 3'-end with an oligodeoxynucleotide. One drawback of this approach is that coupling efficiency decreases with increasing length of the oligonucleotides used.

4.4 Rapid Methods for the Construction of Genes

In this section we describe a couple of methods which in a practical sense can be very useful in the rapid construction of a gene or a set of closely related genes. Closely related does not necessarily mean a mutant but can also refer to different species.

4.4.1 Shot Gun Gene Synthesis

This special type of gene synthesis was made feasible by the recent progress in the chemical synthesis of oligonucleotides, especially by the rapid simultaneous preparation of oligonucleotides by the filter disc method [61], and through the efficient method of dideoxy DNA sequencing [19]. Grundstroem et al. [256] described the synthesis of a high number of SV40 enhancer DNA segments which were assembled from a set of 15 overlapping oligonucleotides and cloned directly into a M13 vector without any purification step prior to transformation of *E. coli*. The synthetic DNA fragment was designed to have the sticky ends complementary to unique restriction enzyme sites of the M13 polylinker. Therefore, only those synthetic DNA molecules which possess both of the chosen 5'- and 3'-sticky ends allow the vector to get circularized and transform *E. coli*. All the other not fully ligated fragments will not give circular closed DNA. Mutant DNA fragments can easily be obtained by replacing the corresponding oligonucleotides within the set through the mutagenic oligonucleotides.

It was found that approximately 50% of the analyzed clones contained the expected sequence. Using this approach in combination with the sequential cloning technique (see 4.2.1.4) Brodin et al. [216] reported the synthesis of a gene encoding the bovine intestinal calcium binding protein from five sets of 11, 3 times 7 and 2 oligonucleotides which were cloned into appropriate M13 derivatives. Two nucleotide alterations were found relative to the designed sequence.

4.4.2 "In Vivtro" Method for Gene Synthesis

This approach makes use of the enzymatic apparatus inside the cell to phosphorylate and join chemically synthesized oligonucleotides. The term "in vivtro" is justified because both "in vitro" and "in vivo" features were combined [257]. The construction of a 120 bp synthetic fragment from 6 oligonucleotides (22 to 54 nucleotides in length) was achieved by directly transforming competent cells with a mixture of these six complementary oligonucleotides and an appropriate linearized plasmid. In the case where the oligonucleotides were introduced as their 5′-phosphates, one out of 100 transformants was positive in screening. When both "in vivo" polynucleotide kinase and "in vivo" DNA ligase activity were exploited by use of unphosphorylated oligonucleotides the expected clone could be found although with a lower efficiency. The practical advantage of the "in vivtro method" compared to the "shot gun" approach is probably of less importance than theoretical considerations.

4.4.3 "Mismatch" Gene Synthesis

In the preceding chapters synthetic duplex DNA was designed such that the upper strand was coding for a polypeptide product whereas the noncoding strand merely

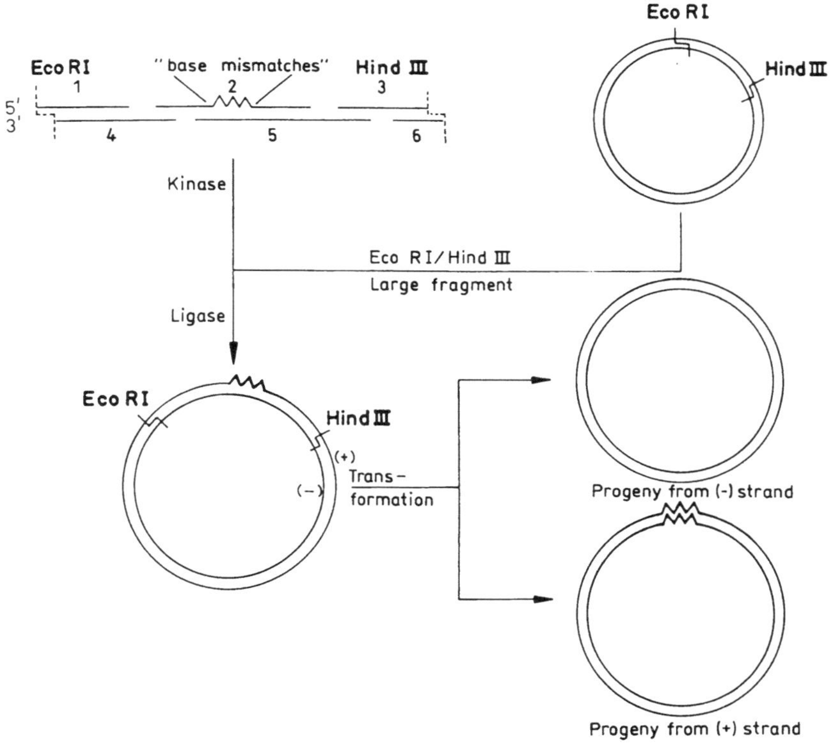

Fig. 17. Strategy for "mismatch gene synthesis" [173]. Synthetic oligonucleotides (1–6) are annealed, kinased and ligated into a cleaved vector. Due to base mismatches transformation results in two different progenies

served the purpose to maintain the duplex structure. In a recent paper, Sung et al. [173] presented a new design where both DNA strands have been utilized to code for two related polypeptides. This was verified by the construction of a DNA heteroduplex with a base "mismatch" in the gene (Fig. 17). After insertion of the "hybrid" or "mismatch" gene into a plasmid, each individual plasmid strand can act as a special template in the DNA replication process which will afford after one cycle two plasmids encoding two different polypeptide products. Since in the original mismatch duplex both strands were made from synthetic oligonucleotides, the repair system of the host cell should repair both strands with approximately the same efficiency thus giving a 1 : 1 ratio of the two different derivatives. For the preparation of a hybrid gene encoding truncated versions of parathyroid hormone (PTH), namely PTH (1–44) and PTH (1–34), eight synthetic oligonucleotides were phosphorylated, annealed to a 3 bp mismatch DNA heteroduplex and ligated directly into an appropriately cleaved vector. After cloning of this material, both gene constructions were identified by colony hybridization and verified by dideoxy sequencing.

Using the same approach the cloning of two genes encoding human and mouse epidermal growth factors (EGF) was accomplished from one synthetic DNA duplex containing 16 single and double base pair mismatches [200]. The sense strand of the 174 bp heteroduplex DNA encoded the human EGF, whereas the nonsense strand indirectly encoded the mouse EGF. The mismatched gene was constructed by ligation of seven overlapping oligonucleotides with a linearized plasmid. After transformation of *E. coli* HB 101 (*rec A* 13) both progenies harbouring either the human or the mouse EGF gene could be found. However, when *E. coli* JM 103 was used as host in transformation, only progenies encoding different chimeric EGF molecules resulting from crossover of the human and mouse EGF gene sequences were obtained.

4.5 Mutagenesis of Synthetic Genes

All known methods for the mutagenesis of nonsynthetic genes can be used for synthetic genes as well. To review in vitro mutagenesis is beyond the scope of this article (for an excellent review see Smith [258]). We will focus on special methods preferably used for synthetic genes.

4.5.1 M13 Mutagenesis

Oligonucleotide-directed site-specific mutagenesis using single-stranded vectors like M13 is an attractive method for both synthetic and natural DNA because a) screening by means of the mutagenic oligonucleotide as hybridization probe indicates the mutant, b) pure single-stranded DNA can be prepared from the packaged phage very easily and c) direct sequencing using the single-stranded DNA as template in the Sanger chain termination method is straightforward. The basic principle of M13 mutagenesis has been outlined in Sect. 4.3.2. If one oligonucleotide is used as primer this variant is called "the single primer method" [247]. *E. coli* owns a repair system [259] which favourably repairs non-methylated DNA. Since synthetic DNA is not methylated, yields in mutants are much lower than in theory. The "two primer method" [260] gives improved results by suppressing the exonuclease activity, the latter can lower the efficiency. Significantly improved mutation efficiencies of more than 70% have been reported for the "gapped duplex method" [249, 261]. The gapped duplex DNA which presents only the DNA insert to be mutagenized in a single-stranded form, constitutes a non-amber M13 mutant. It is transformed in a non-suppressor host to select for the mutant. Another method of strand selection was presented by Carter et al. [262] allowing repeated rounds of mutagenesis without need of recloning by use of the related EcoK and EcoB sites differing only in one nucleotide change. The Kunkel method [263] makes use of a deficiency in dUTPase and uracil glycosylase thus allowing partial incorporation of deoxyuridine in place of thymidine. The "phosphorothioate method" developed by Eckstein [264] is based on the fact that certain restriction enzymes as NciI or PstI cannot cleave phosphorothioate DNA. Therefore the simple one primer method is used in the initial step except that primer extension is done using addition of α-S-dCTP. After nicking the natural DNA strand by NciI, exonuclease digestion of the wildtype strand, repolymerisation and transformation in *E. coli*, mutation efficiencies of up to 95% are reached.

4.5.2 Cassette Mutagenesis

When dealing with synthetic genes this is probably the most useful and convenient strategy for site-specific mutagenesis, since synthetic genes can be specifically designed for this purpose. Cassette mutagenesis involves the replacement of a DNA fragment which is flanked by unique restriction enzyme cleavage sites with a synthetic mutagenic DNA duplex. In other words, this process basically is a partial gene synthesis. Genes coding for ribonuclease S protein [206], calmodulin [217], bovine rhodopsin [141] and growth hormone releasing hormone (GHRH) [174] were designed especially for site-specific mutagenis by placing numerous restriction enzyme cleavage sites at short

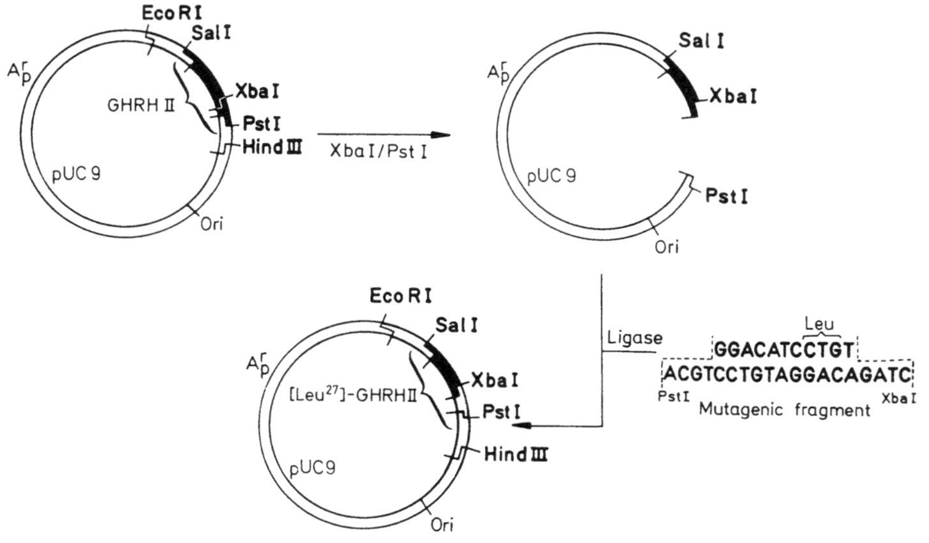

Fig. 18. Cassette mutagenesis. Mutation of GHRH II gene [174] to [Leu^{27}]-GHRH II

intervals within the gene. In the case of GHRH-Gly^{45} the codon for methionine 27 had to be replaced by a codon for leucine to allow cyanogenbromide cleavage of a fusion protein with a part of galactosidase [174]. This could be achieved by cleavage of the synthetic gene with the unique restriction enzymes PstI and XbaI and religation with the appropriate synthetic duplex (Fig. 18). Analogously, by inserting a set of mutagenic DNA duplexes one can generate a set of mutants in one experiment. Then, using an appropriate screening system only the most interesting, e.g. the most active, mutant is selected. In a targeted random mutagenesis experiment [265] ambiguously synthesized oligonucleotides were used to mutagenize sequences immediately 5′ of an ATG initiation codon of the bovine growth hormone gene. Cassette mutagenesis has also been successfully applied to generate 19 amino acid substitutions at position 222 in the subtilisin protein sequence [266].

There are two variations of the cassette mutagenesis method having the advantage that only one synthetic oligonucleotide instead of a duplex consisting of two oligonucleotides is needed. First, it is possible to construct mutants by ligating a synthetic oligonucleotide into a precisely preformed single-stranded gap covering the region to be mutagenized of an otherwise double-stranded gene [267]. Second, single-stranded oligonucleotides bearing ends which are complementary to those left by restriction enzymes giving 3′- and 5′-protruding ends (see also 4.3.1) can be ligated directly into appropriate cleaved vectors [245, 246].

4.5.3 "Shot Gun" Ligation

The principle of this technique has been discussed in Sect. 4.4.1. It surely is more complex than cassette mutagenesis (partial gene synthesis) and it needs more expertise to construct the total gene each time from scratch. But it has the advantage that mutagenesis experiments can be done without need of any internal restriction enzyme cleavage sites. Combined with the "mismatch" gene synthesis approach two or more mutants can be created at a time.

5 Design Considerations of Synthetic Genes

5.1 Selection of the Nucleotide Sequence of a Gene

Once the amino acid sequence of a protein from peptide sequencing or from the corresponding cDNA is known the question arises what concrete nucleotide sequence should be chosen for the synthetic gene within the limitations of the genetic code. This is the most crucial point in gene synthesis since all the considerations discussed in the different sections of this article must be reduced to a common denominator starting from the chemistry we use, the mode of gene construction, the type of vector system and host for expression of the gene up to basic secondary structure considerations. Taking into account all the different restrictions it is obvious that at some point compromises have to be made to get down to the final sequence.

5.1.1 Codon Usage

Since coding sequences in natural DNA do not use synonymous codons with equal frequencies and the pattern of nonrandom codon usage varies with the level of gene expression it apparently makes sense to use a subset of "optimal" codons for the construction of synthetic genes. And this has been done for almost all synthetic genes which so far were designed for the expression in *E. coli*.

By investigation of the codon usage of highly expressed proteins in *E. coli* rules were deduced which allowed the understanding of the strikingly different codon usage in highly expressed bacterial mRNAs [15, 268, 269]. The systematic tabulation of codon usage deduced from Gen Bank sequence data has been extended to many other organisms [270]. Although the codon usage has been exploited for most synthetic genes designed for expression in *E. coli* only few hard data have been published concerning this subject. Direct experimental evidence, of course, can not easily be presented, because a change in codon usage somehow always influences the secondary structure of a mRNA. In the case of chloramphenicol acetyltransferase (CAT) which is highly expressed in *E. coli* it was shown that unfavourable codon usage can significantly affect the rates of protein synthesis [271]. But this effect could be detected only at very high transcription rates. The importance of codon usage seems also to differ from host to host. Although codon usage seems to play a role in gene expression in prokaryotic hosts, experiments using higher cells rather suggest the opposite. It has been reported that a synthetic gene for human growth hormone [272] can be expressed efficiently in yeast despite the fact that the gene was designed with codon preference for *E. coli*. Similarly, the results obtained in expression of the synthetic interferon-α_1 gene (*E. coli* codon usage) in mammalian cells argue against a crucial role of codon usage in higher cells [273]. Furthermore, unusually high-level expression of a foreign gene, namely hepatitis B virus core antigen, in *Saccharomyces cerevisiae* has been found [274], although the analysis of the coding sequence of this gene revealed a rather low index of codon bias for *S. cerevisiae*. On the other hand, comparison of 110 known yeast gene sequences clearly revealed a codon usage bias in yeast [275]. Chemical synthesis of genes using yeast-preferred codons and their expression in yeast have been performed including human EGF [175], human ANP [176] and connective tissue activating peptide-III [199], but experimental data are not conclusive regarding the codon usage. Some species like *Cellumonas fimi* [276] show an extreme bias (98%) for G or C in the third position of the codons. Therefore, codon usage may be more important for those hosts.

5.1.2 Restriction Enzyme Sites

One of the greatest advantages of synthetic genes is the possibility to introduce numerous unique restriction enzyme sites throughout the sequence. This, on one side, is very helpful in the construction of genes as it permits the sequential cloning of gene fragments (see 4.2.1.4) and, on the other hand, allows for subsequent specific alteration of a gene (see 4.5.2). Usually restriction enzyme sites producing on cleavage sticky ends are preferred over those creating blunt ends. For cloning of genes or subcloning of gene fragments it is convenient to choose sites which are present in commercially available plasmids. The polylinker containing pUC vectors [277] and M13mp [278] vectors are favourably used for this purpose. Terminal restriction enzyme sites are usually preformed as 5′- or 3′-protruding ends to facilitate direct integration into appropriately cleaved plasmids. But synthesis of blunt ended fragments, and subsequent cleavage with suitable restriction enzymes works as well.

For sequential cloning of gene fragments there is frequently no site available which is compatible with commercial cloning vectors. Then use of temporary restriction enzyme sites is indicated. One possibility is to extend the gene for several nucleotides to create a clonable site. After amplification of the gene fragment the desired sequence can be generated by cleavage with an enzyme whose recognition site is juxtaposed to the temporary site. Another valuable approach is to insert into a sequence a couple of nucleotides so as to allow cloning. On a later stage of synthesis the additional nucleotides are removed by nuclease S1 [238]. The insertion points can be chosen more or less arbitrarily at any place of a DNA sequence containing a pair of selfcomplementary dinucleotides (AT, TA, GC or CG). For instance, a temporary site for Eco RI (GAATTC) is generated by insertion of a AATT tetranucleotide at a GC dinucleotide sequence (Fig. 19). After cloning of the plasmid containing the gene fragments, this can be recleaved by EcoRI. The protruding tetranucleotides AATT are removed by nuclease S1 cleavage and the formed blunt ends are religated to give the desired sequence.

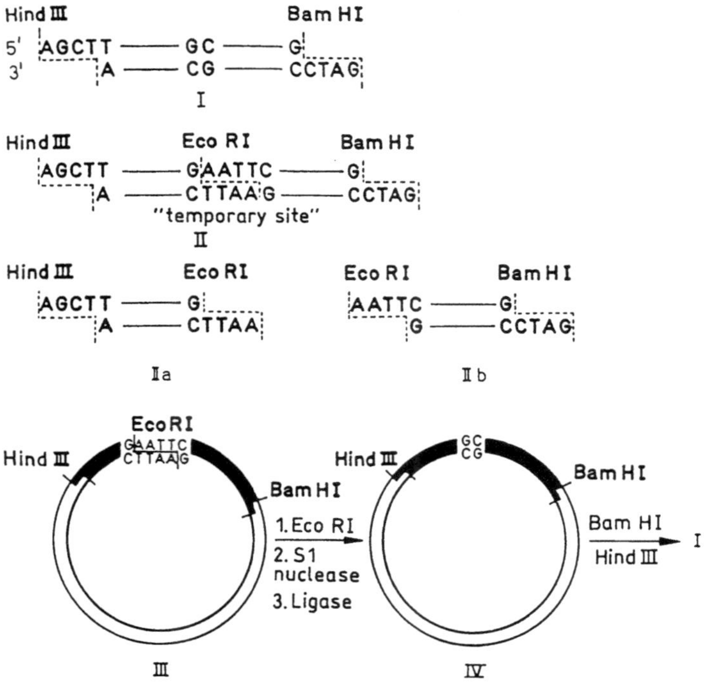

Fig. 19. Temporary restriction enzyme cleavage site [238]. I: sequence to be synthesized, II: the same sequence as I, but with a temporary Eco RI site inserted, IIa, IIb: DNA fragments to be synthesized, III: plasmid with target sequence plus additional temporary restriction enzyme cleavage site, IV: plasmid harbouring the target sequence I (Eco RI temporary site removed)

In praxi it turned out to be helpful to have a unique restriction site just downstream of the ATG start codon and another just upstream of the translation stop codon. This facilitates rapid shifting of the gene into different expression systems by means of suitable adapters. Another approach is to place type II restriction sites at both ends of the gene [175, 201] in a position where on cleavage with this enzyme followed by nibbling back the generated protruding ends with the single-strand specific nuclease S1 the entire gene without start and stop codon is accessible. By virtue of appropriate linkers the gene can be inserted into a new expression system. In such cases, where a specific restriction enzyme site is not wanted during the construction of a gene but rather for subsequent modification of this gene, the recognition site for this enzyme can be preformed such that on changing one nucleotide by site-specific mutagenesis the site becomes apparent. Enzym recognition sites are tabulated in Ref. [279, 280].

5.1.3 Secondary Structure Considerations

Once a gene sequence has been optimized with respect to codon usage and unique restriction enzyme sites, it usually has to be subdivided into oligonucleotides of a size being accessible by chemical synthesis. Then each oligonucleotide must be checked for intra or intermolecular secondary structure problems, especially, if the G-C content is high. Already in the pioneering work of Khorana and coworkers [159] disturbing self-complementarity (Fig. 20a), "self-structures" (Fig. 20b) and hairpin structures of oligonucleotides (Fig. 20c) were discussed. The more oligonucleotides are used in

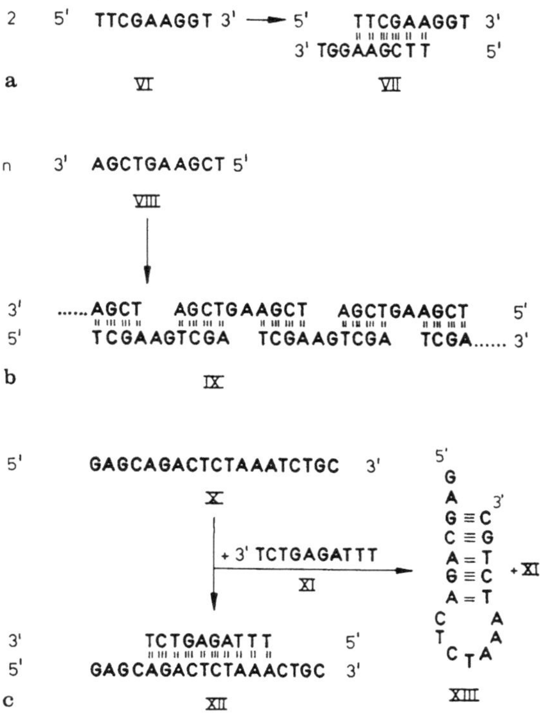

Fig. 20a–c. Secondary structure problems [159]. **a** Selfcomplementarity of oligonucleotide VI may result in duplex VII. **b** IX is a possible "self-structure" of the oligonucleotide VIII. **c** Annealing of oligonucleotides X and XI yields the desired duplex XII, alternatively the hairpin-structure XIII may be formed

a one-pot ligation reaction the more complex is the secondary structure computation, because each oligonucleotide must be investigated for unwanted intramolecular or intermolecular interactions with itself or with all the other ligation partners. This can only be done with the help of appropriate computer facilities. If longer synthetic oligonucleotides are used, duplex structures can be preformed by pairwise hybridization of the partially complementary oligonucleotides. For further ligation mainly the interaction of the protruding ends must be analyzed. In this respect very long synthetic oligonucleotides representing the coding and noncoding strand of a gene or gene fragment which are hybridized and directly incorporated into a cloning vector [243)] cause least trouble. In contrast, the "fill-in" method in combination with long oligonucleotides requires very careful investigation of potential secondary structures. Calculation of secondary structures of RNA will be discussed in Sect. 5.3.2.3, the methods being applicable to oligodeoxynucleotides as well.

5.2 Consideration of Expression Mode of a Gene

5.2.1 Direct Expression

This is the most straightforward mode of expression in prokaryotic hosts whenever the expressed protein is a stable and, for the host, nontoxic product. The synthetic gene is placed directly under the control of an inducible [187)] or a constitutive [281)] transcriptional-translational control unit. For the expression of the human leukocyte interferon gene [232)] the coding region of the synthetic gene was preceded by an ATG start codon plus the protruding ends of a BamHI site and followed by a stop codon and the protruding ends of a SalI site [169)] (Fig. 21). The two single-stranded cohesive ends permit the ligation of the gene in a predetermined orientation into a plasmid under the transcriptional control of the lac promotor. The consequence is that the product is expressed as a methionyl-IFN-α. In most cases the additional Met is tolerable regarding biological activity and antigenicity, but may be removed by cyanogenbromide cleavage or by specific proteolytic processing. Direct expression of polypeptides up to a size of 50 to 80 amino acid residues usually is not successful (except urogastron, 53 amino acids) [181)] and other modes of expression have to be considered.

5.2.2 Expression of a Fusion Protein

In 1977, the first successful synthesis of a functional oligopeptide product from a chemically synthesized gene was reported [166)]. In this early work it already became apparent that small proteins, like the tetradecapeptide somatostatin, have to be stabilized by fusion to a protein to avoid proteolytic degradation. For instance, pSOMI (Fig. 22), designed for direct expression of somatostatin under control of the lac PO unit, did not give the desired peptide. Integration of the β-galactosidase gene just upstream to the somatostatin gene was expected to lead to a fusion protein when expressed under lac PO control. Chimeric proteins consisting of ten amino acids of β-galactosidase and somatostatin could not be successfully expressed whereas a fusion protein consisting of virtually the entire β-galactosidase and somatostatin resulted in the expression of a stable precursor. The fusion protein was designed to have a Met preceding the somatostatin part which allowed in vitro cleavage with

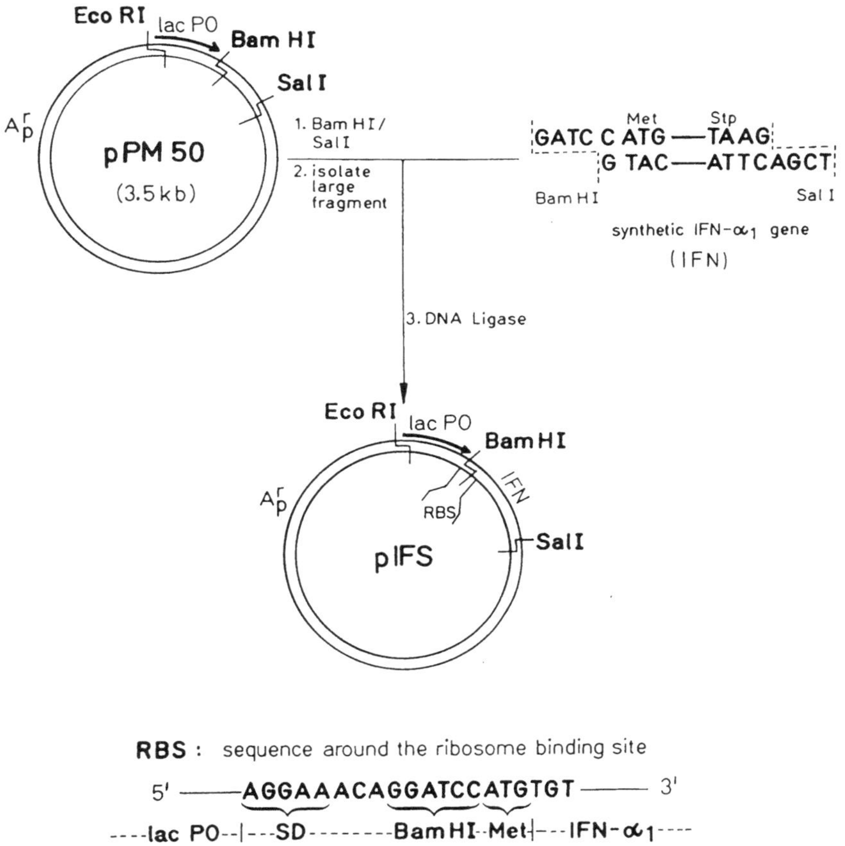

Fig. 21. Construction of recombinant plasmids for the direct expression of IFN-α_1 [169, 232]. (SD = Shine-Dalgarno sequence, lac PO = lac promotor-operator control unit, Stp = stop codon)

cyanogen bromide to give biologically active somatostatin. If a polypeptide to be expressed contains an internal Met, cyanogen bromide cleavage of the precursor is no longer possible. One solution of this problem is to mutagenize the product within the limitations of hydrophobicity [282] and secondary structure predictions [283]. For example, biological activity was retained for Leu^{27}-growth hormone releasing hormone [174]. Another possibility is to incorporate an alternative processing signal N-terminal to the desired polypeptide. Asp-Pro dipeptide sequences are susceptible to acid hydrolysis [284, 285]. Dibasic amino acid sequences as Arg Arg are hydrolysed by trypsin/carboxypeptidase B [286]. Several specific endoproteases were utilized as for instance blood coagulation factor Xa [240] for cleaving fusion proteins.

One problem of this approach is that about 98% of the expressed fusion protein is discarded and what is more, β-galactosidase contains 23 Met residues which are also cleaved with cyanogen bromide to give small peptide fragments which render the purification of the desired product more difficult (Fig. 22). Therefore, shortened versions of β-galactosidase [287] consisting of 590 amino acids instead of the original

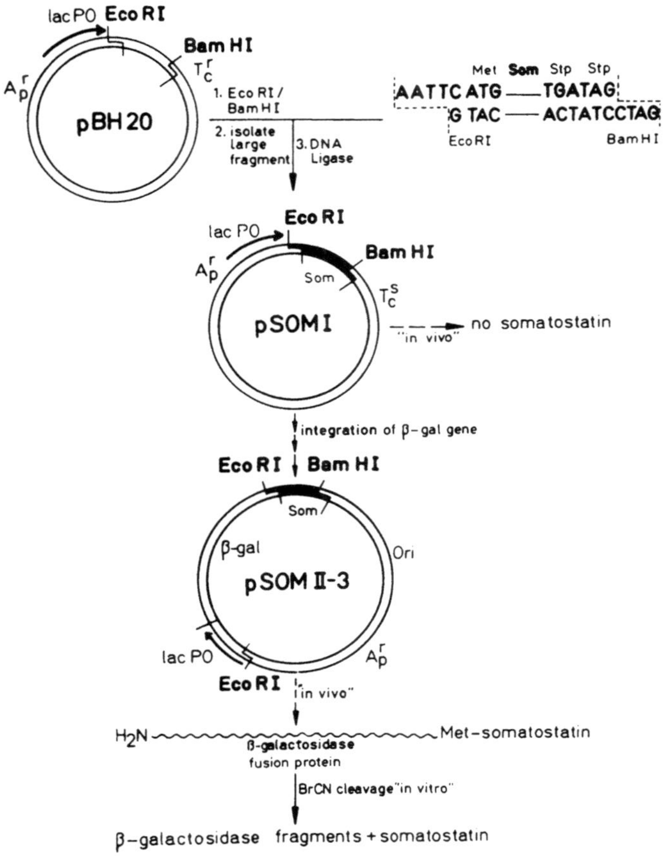

Fig. 22. Construction of a plasmid for the expression of somatostatin as a fusion protein with β-galactosidase [166]. (som = somatostatin coding region, β-gal = β-galactosidase coding region)

1007 amino acids were constructed which gave improved yields. Similarly, shorter proteins still forming inclusion bodies in *E. coli* and thus stabilizing the protein against degradation were used, e.g. parts of Trp E [177], interferon-γ [178] or human growth hormone [209]. Furthermore, enhanced proinsulin expression could be achieved by addition of a short homooligopeptide to the amino terminus of proinsulin [288]. Of course, also ternary constructions are possible where the protein of interest is hidden between two heterologous protein parts.

5.2.3 Transport Expression

A decisive disadvantage of the expression modes discussed in the two preceding chapters is that the desired peptide products frequently are not obtained directly in their biologically active form. For instance, disulfide bridges are not or only incorrectly tied up or the active structure is destroyed during cleavage of the fusion protein under denaturing conditions. One possibility to overcome the problems in protein folding, disulfide bond formation and proteolytic degradation is to direct the nascent polypeptide chain through the cytoplasmic membrane out of the reducing environment present in the cytoplasma of bacteria by means of bacterial or other leader sequences. In *E. coli*, translocation was described mainly to the periplasmic space, except for a protein A fragment construct with insulin-like growth factor I [289] which was efficiently secreted to the culture medium of *E. coli*. Synthetic genes for human EGF [179] and hirudin [290] as well as a gene for growth hormone [291] were fused to the signal sequence of the *E. coli* alkaline phosphatase and expressed. In all instances the biologically active products could be found in the periplasmic fraction which suggests that signal peptidase cleavage according to the signal hypothesis [292] had occurred. In the case of hirudin two forms of the processed protein were obtained. Furthermore, a chemically synthesized alkaline phosphatase signal sequence DNA fragment, which was designed as a movable element, was used for secretion of insulin in *E. coli* [293].

Generally higher yields of products are obtained by excretion in hosts other than *E. coli*, e.g. *Staphylococci* [294] or *B. subtilis* [295] where the product is recovered simply from the medium. Expression and protein secretion in yeast now seems to gain more and more importance, especially in combination with the prepro-α-factor leader region [296, 297]. Only recently, the expression of synthetic genes encoding connective tissue activating peptide-III [199] and human atrial natriuretic peptide [176] using the α-factor system have been reported. In another paper [207] the expression of a synthetic human lysozyme gene was reported, which was secreted and processed in yeast using a synthetic chicken lysozyme signal sequence. For more information about the general features of signal peptides we refer to special reviews [298, 299].

5.2.4 Multiple Gene Approach

Two basic strategies have to be differentiated using the multiple gene approach. First, the polycistronic tandem gene systems, which exploit the gene dosage effect and yield the protein in a form identical to the direct expression mode. Second, tandemly linked genes, which on expression result in a fusion protein consisting of more or less identical domains, are more frequently used although a posttranslational processing step is required (Fig. 23b).

The first concept is borrowed from bacterial operons which often are polycistronic (Fig. 23a), that means a series of different, but related genes are transcribed into one mRNA from their common transcriptional control unit, examples being the lactose operon and tryptophane operon. The method was applied to the high level expression of human leucocyte interferon [300]. Up to four interferon genes, each having a suitable upstream ribosome binding site, were combined via intercistronic spacers and expressed under the control of one promoter and one terminator. The amount of expressed interferon was found to be directly proportional to the number of interferon genes in the corresponding plasmids. Interestingly, high stability of the tandem plasmids in a rec A host (for more than a hundred generations) was reported. Moderate expression levels of proinsulin were reported for a construction involving two proinsulin genes connected by a short intercistronic spacer [182]. The second strategy was applied for the expression of proinsulin [301]. So far, no high level direct expression of human proinsulin has been reported due to the fact, that the half-life of this polypeptide in *E. coli* is 2 min. However, when multiple copies of the proinsulin gene are

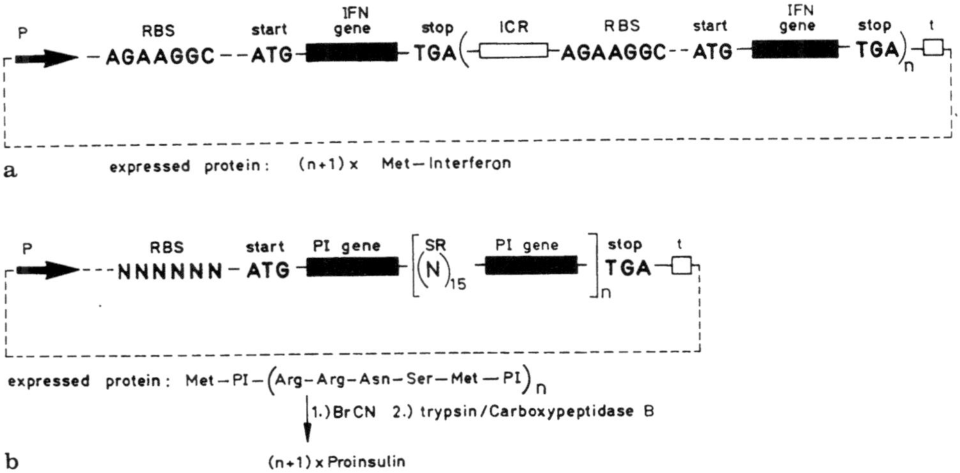

Fig. 23a and b. Multiple gene approach constructions. **a** Polycistronic tandem gene construction as applied for leukocyte interferon [300]. Up to four IFN genes (n = 0–3) were tandemly arranged. **b** Multi-domain tandem gene construction as applied for proinsulin [301]. Up to five proinsulin genes were fused tandemly (n = 0–4). (P = promotor, RBS = ribosome binding site, ICR = intercistronic region, t = transcription terminator). (PI = proinsulin, SR = spacer region coding for Arg Arg Asn Ser Met)

linked so as to give a fusion product consisting of multiple proinsulin domains, the stability of the product is significantly increased. There are three tandem-linked proinsulin domains necessary for stabilization and up to five (or seven with an additional 80 amino acid lac leader sequence) joined proinsulin coding sequences were investigated. Maximum expression was achieved with 3 to 4 joined sequences, whereas 7 joined genes resulted in a lower yield in protein. As a spacer in-between the proinsulin sequences Arg-Arg-Asn-Ser-Met was used and it can be removed by a cyanogen bromide cleavage followed by digestion with trypsin and carboxypeptidase B. The reason for stabilization of proinsulin in the multicopy arrangement is the same as for the β-galactosidase fusion proteins, namely the formation of inclusion-like bodies which can be observed by electron microscopy. The multicopy gene approach has also been described for substance P [211], growth hormone releasing hormone [180] and Met-enkephalin [212]. Already in 1980, the expression in *E. coli* of synthetic repeating polymeric genes (up to 150 repeats) for L-aspartyl-L-phenylalanin was presented as a method in the fermentative production of aspartame [222].

5.2.5 Genes Designed for Posttranslational Modification of the Expressed Protein

This subject is a direct consequence of the fusion protein technique as well as the non-polycistronic multiple gene approach and many other applications. One of the first examples of posttranslational processing was the separate expression of the insulin A and B chain, which were posttranslationally combined to give the active insulin [168]. A more sophisticated approach was the construction of a gene encoding a human proinsulin analogue, where the normal 35 amino acid connecting peptide was replaced by a sequence of only six amino acids [172]. This mini-C-peptide sequence Arg-Arg-

Gly-Ser-Lys-Arg contained the signal for posttranslational proteolytic processing. Very recently, a similar approach was used for the expression of insulin in yeast [302], but this time posttranslational processing involved a transpeptidation reaction. This is one of the rare examples of semisynthesis of peptides using recombinant material which presumably will become an important technique in the future. To date, for the introduction of D-amino acids into polypeptides produced by genetic engineering techniques, semisynthesis is the method of choice.

Since many important peptides, mostly containing less than 50 amino acids, occur in nature as a C-terminal carboxamide which is not accessible in prokaryotic hosts, again a posttranslational modification reaction is needed. Synthesis and expression of a gene for [Leu^{27}]-GHRH-Gly^{45} has been reported [174] which involves two changes in the sequence compared to the natural GHRH-carboxamide. Both changes were designed on the DNA level to allow posttranslational modification of the expressed peptide product. First, Met^{27} in the natural sequence was replaced by Leu, considering hydrophobicity [282] as well as secondary structure predictions [283] to facilitate posttranslational cyanogen bromide cleavage. Second, a Gly was added at the carboxy terminus of the sequence which by an amidating enzyme [303] allows the conversion of the precursor to the desired GHRH-amide [228].

A further solution to this problem has been described yielding homoserine amides [180]. A precursor of GHRH where the natural sequence is flanked by a Met on each side and the internal Met^{27} is replaced by Leu was subjected to cyanogen bromide cleavage. The resulting homoserine (Hse) lactone derivative [Leu^{27}, Hse^{44}]-GHRH-(1–44)lactone could be transformed by ammonia treatment to [Leu^{27}, Hse^{44}]-GHRH-NH_2. In addition to these examples all recognition sequences for proteolytic or chemical cleavage can be programmed in a gene (see 5.2.2).

5.3 Synthetic DNA Fragments Controlling Gene Expression

Of course, chemical DNA synthesis is not limited to the synthesis of genes, but can also be applied to the synthesis of transcriptional and translational control units which in turn give rise to expression of the synthetic gene. Thus, the synthesis of a gene control region involving an idealized promotor, an operator, and a ribosome binding site and its application to the expression of a synthetic IFN-γ gene was reported [305]. There are many investigations involving synthetic promotors, operators, terminators, and synthetic ribosome binding sites. For didactic reasons the following section is systematically subdivided, but this should not imply that any of these aspects can be considered independently from all the others to optimize gene expression. On the contrary, the interplay of all the discussed factors finally determines the results. Sometimes investigations are misleading because one aspect is discussed out of the context.

5.3.1 Control of Transcription

5.3.1.1 Synthetic Promotors

A promotor is a special sequence of a DNA molecule usually in the 5′-untranslated region which is important for transcription of a gene by interacting with RNA poly-

```
BamHI

GGATCCTAAATAAATTCTTGACATTTTTTAAATAATTTGGTATAATGTGTG
CCTAGGATTTATTTAAGAACTGTAAAAAATTTATTAAACCATATTACACAC
                 ________                    ________
                   -35                          -10

                             PROMOTOR


                                                  EcoRI

GAATTGTGAGCGGATAACAATTTCACTTTAAAGAGGATCTAGAATTCATG
CTTAACACTCGCCTATTGTTAAAGTGAAATTTCTCCTAGATCTTAAGTAC
                                ______            ___

Lac OPERATOR                      SD            START
```

Fig. 24. Synthetic control region consisting of the promotor, operator and Shine-Dalgarno (SD) sequence [305]

merase. For prokaryotic promotors a lot of sequence information is available which was used to deduce general features of these promotors [304]. These are the —35 region (RNA polymerase binding site) and the —10 region (Pribnow box) which are connected by an AT-rich spacing sequence (Fig. 24). The construction of several synthetic promotors has been reported including a modified *E. coli* promotor [306], a model promotor [307], a consensus sequence promotor [239], a consensus sequence promotor as part of a synthetic control region (Fig. 24) [305], a bacteriophage lambda P_R promotor [308], a T5P25 promotor [281, 309] and a trp promotor [310]. Although the individual investigators found a high efficiency of the synthetic consensus promotors, a systematic investigation of the in vivo strength of 14 promotors revealed that, in the system applied, one of the synthetic consensus promotors belonged to the less efficient signals and that information located in the transcribed region could influence promotor strength as well [311].

5.3.1.2 Synthetic Operators

Among the promotors mainly used to date we have to differentiate between two types. Both the constitutive type and the inducible type were successfully applied for gene expression. The promotors under control contain in addition to the sequences discussed in 5.3.1.1 an operator sequence. The operator sequence can overlap with the promotor sequence (e.g. trp operator) or is a separate element (e.g. lac operator). Usually, regulation is effected by the limited availability of a low molecular weight molecule (e.g. phosphate for alkaline phosphatase), by addition of a low molecular weight inducer (e.g. IPTG for the lac system), by heat induction (e.g. lambda P_L) or even by light. The best investigated system is the lac operon, which is under negative control [312]. The start of transcription is inhibited by the binding of a lac repressor protein to the lac operator DNA sequence. Induction of the promotor can be achieved by addition of isopropylthiogalactoside which turns the repressor into a non operator-binding form. Synthetic idealized lac operators with a perfect dyad axis of symmetry compared to the imperfect wild type operator were constructed [313, 314]. These perfectly symmetric operators bind lac repressor up to 8-fold more tightly than the natural operator both in vivo and in vitro. Since the wild-type lac promotor is not completely shut off by the wild type operator, a stronger promotor in the same arrangement will afford an adequate operator in order not to render the system more leaky [315]. The synthetic lac operator was also used for the construction of an inducible rRNA promotor P_2 which is otherwise constitutive [316]. However, repression is incomplete when an additional rRNA promotor P_1 is fused upstream from P_2 in a tandem construction. Of interest is also, that the synthetic lac operator mediates repression through the lac repressor not only downstream from the lac promotor (Fig. 24) but also when introduced upstream from the lac promotor [317].

5.3.1.3 *Synthetic Terminators*

In the preceding two sections the start of transcription using synthetic DNA has been discussed as an important factor in gene expression. On the other hand, transcription termination is of significance too, especially, when dealing with strong promotors. Thus, it was shown that insertion of a strong terminator derived from bacteriophage T4 behind the IFN-γ gene of an expression plasmid increased the yields of IFN-γ by a factor of two [318]. Enhancement of the overall yield in expression of the urogastrone gene by a factor of 5–10 by insertion of a transcription terminator from bacteriophage fd has been found [319]. Consequently, a gene for an intestinal calcium binding protein having a synthetic rrnC transcription terminator downstream the translation stop codons was constructed [216] (Fig. 25). Furthermore, terminator structure may be idealized or two or more transcription terminators can be tandemly fused. Of course, internal transcription terminator-like sequences should be avoided within a gene [188].

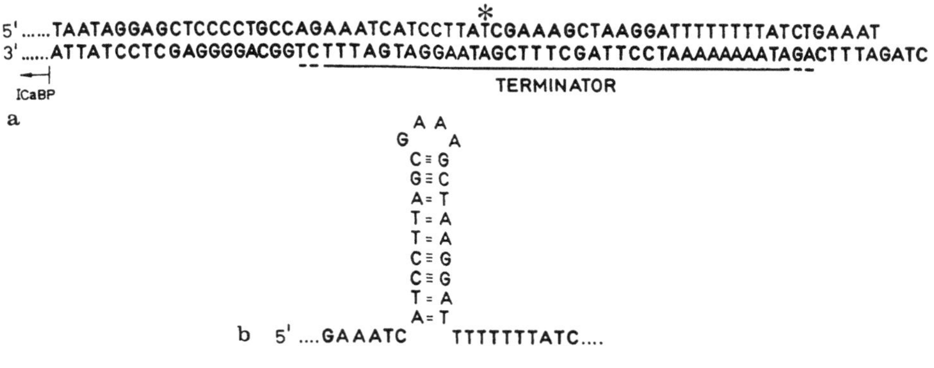

Fig. 25 a and b. Synthetic terminator of transcription. **a** Nucleotide sequence of the 3'-nontranslated region of the bovine intestinal calcium binding protein (ICaBP) involving the synthetic rrnC transcription terminator [216]. **b** Possible secondary structure of the natural rrnC terminator [334]. Asterisk marks a nucleotide exchange between natural and synthetic sequence

5.3.2 Control of Translation

Once a DNA has been efficiently transcribed into the corresponding mRNA, the latter must have a halflive time long enough to guarantee translation into a protein. For successful translation the mRNA has to have an appropriate ribosome binding site (RBS) as well as a signal for translation initiation and termination.

5.3.2.1 Synthetic Ribosome Binding Sites

Biosynthesis of proteins is initiated by binding of the 3'-end of 16S RNA via base pairing to a complementary sequence of the mRNA. This ribosome binding site (RBS) was shown by Shine and Dalgarno [320] to have in prokaryotic hosts the consensus sequence AAGGA. This so-called Shine-Dalgarno (SD) sequence is generally 4–9 bases long and is located 3 to 12 nucleotides upstream of the translation initiation codon.

Genes originating from eukaryotic sources do not possess a regulatory signal to be recognized by prokaryotic ribosomes. This fact has led to construction of expression systems using synthetic RBS's. Efficient synthesis of SV40 t antigen using a synthetic RBS has been reported [321]. Varying spacing between the RBS and the start codon has been investigated [322]. Synthetic ribosome binding sites have been applied furthermore for the expression of interferon-α gene fragments [323], urogastron [319], IFN-γ [305, 318], IFN-β [324], interleukin-2 [325] and many other proteins. In several cases the RBS's were optimized regarding the spacing between the RBS and the start codon and considering the secondary structure of the mRNA. Again, optimization is only possible by investigation of the RBS with the context in which it is studied.

5.3.2.2 Initiation and Termination of Translation

The initiation complex consists of 16S rRNA (and initiation factors) interacting with the SD-sequence and of N-formyl methionyl-tRNA interacting with the translation start codon AUG of the mRNA. In *E. coli* the alternative start codon GUG is observed only rarely and UUG was never found. Translation termination is encoded by UAA, UAG and UGA, the latter being not so frequently used in prokaryotes. To avoid in any event overriding of the stop codon synthetic genes often contain two different stop codons in tandem arrangement. The termination codon UGA seems to be used in very rare cases for seleno-cysteine in eukaryotic mRNA [326].

A crucial problem concerning translation initiation is the occurrence of so-called restarts. Restart of translation can take place when an internal AUG codon is preceded by a SD- or SD-like sequence with appropriate spacing of 3 to about 20 nucleotides and results in shortened versions of the expressed protein. Contamination of the desired full length protein product by N-terminal truncated products can be substantial [327, 328]. To circumvent the restart phenomenon potential RBS's just upstream of internal Met codons should be avoided by changing the corresponding codons within the limitations of the genetic code. Sometimes this causes problems for Glu- or Lys-rich sequences, because these aminoacids are encoded by GAPu or AAPu, respectively.

5.3.2.3 mRNA Structure Considerations

It has been observed that a synthetic gene when fused to different transcriptional/translational control units gives considerably higher yields in expression for one construction than for the other. But when a second gene is tested in both constructions just the opposite results are obtained. That means, the efficiency of expression of a gene does not merely depend on the structure of the gene and the strength of the promotor but on the entire context in which the gene is investigated. On one side, the mRNA must fold into a relatively stable secondary structure to survive the time span between transcription and translation. On the other hand, translation only efficiently proceeds from the single-stranded form. Therefore, it seems to be important that the RBS is easily accessible by the 16S rRNA and not hidden in a stem. The same is true for the AUG start codon which should be located in the loop and not in the stem of a hairpin. But for translation initiation it is not essential that the AUG codon resides in a single stranded region [327].

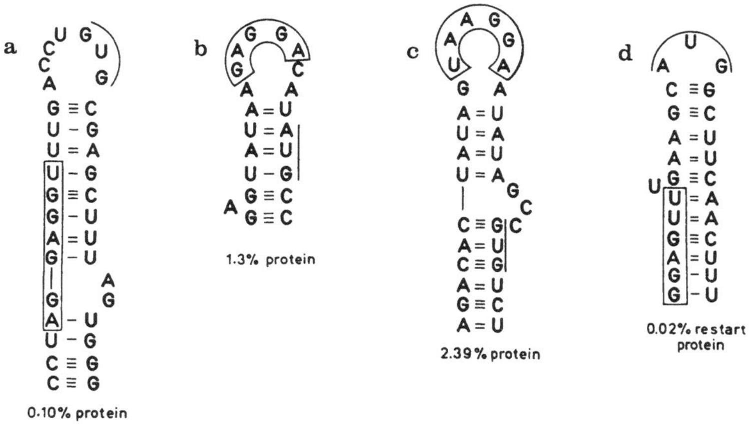

Fig. 26a–d. Secondary structure analysis of RBS regions [328]. Correlation with synthesis of lac Z fusion protein. **a** Bacteriophage MS 2 RBS; **b** bacteriophage Qβ RBS; **c** elongation factor EF-TU A RBS; **d** restart in bacteriophage MS 2 C construction (RBS = ribosome binding site)

Numerous publications appeared on this subject and each new expression construction results in a new mRNA showing an individual secondary structure with its own merits or problems. For example, in the case of IFN-γ expression levels varied up to 50-fold among different constructions. Thus, changing one C into an U at position 11 downstream the start codon increased expression level by a factor of 5. The results could be explained on the basis of mRNA secondary structure calculations [329]. Stem-loop structures that block translation initiation (high $-\Delta G$ values) drastically decreased translation initiation. In another study several heterologous RBS's were fused to the lac Z gene of *E. coli.* There was an over 100-fold variation in expression levels as evaluated by an immuno precipitation method [328]. A correlation between the efficiency and the secondary structure of the RBS's was suggested. Easily accessible RBS's gave higher yields in protein than RBS's involved in stem-formation (Fig. 26a–c). An additional "false" translation initiation site (Fig. 26d) was envisaged to explain a very low expression level of a putatively effective RBS. One problem of a number of theoretical considerations is that only parts of the mRNA structure are calculated, but the remainder sequence might dictate formation of a different overall structure. For computation of the mRNA secondary structures several algorithms were described [330–332]. Three special issues of Nucleic Acids Research were devoted to this subject and other computational methods concerning nucleic acids [333].

6 Future Outlook

Oligonucleotide synthesis has made impressive steps forwards by improving yields in recent years. Thus longer, purer sequences will be the building blocks for synthetic genes making the application of the strategies discussed even more elegant. The segmental approach, when successfully automatized, can cut down on cost and waste. Certainly, besides automation of synthesis, automation of purification or use of robotics [335] is along the lines of future developments. Synthetic genes will also lead to construction of completely new peptides [336] and proteins [208] which can have interesting and new structures and functions and they can help in elucidating this still open and challenging enigma of protein structure function correlation.

We shall witness as a result of recombinant DNA techniques based on synthetic genes the production not only of hitherto unknown peptides but also of proteines, enzymes, receptors and all proteins of living matter which of course leads to the question of ethics and of what we can do and should not do.

7 Acknowledgements

We acknowledge the BMFT, DFG and Hoechst AG for financial support and we thank Helga Zimmermann for the drawings and Evelyn Groeneveld for preparing the manuscript.

8 Abbreviations

aa	amino acid
AMV	avian mycloblastosis virus
Ap	ampicillin
B	nucleotide base
bp	base pairs
cDNA	complementary DNA
CPG	controlled pore glass
DBU	1.5-diazabicyclo-[5.4.0]-undec-5-ene
DCC	dicyclohexylcarbodiimide
DMT	4,4′-dimethoxytrityl
dNTP	deoxynucleotide triphosphate
β-gal	β-galactosidase
HPLC	high performance liquid chromatography
IFN	interferon
IPTG	isopropyl-β-D-thiogalactopyranoside
kb	kilo base(s)
lac PO	lac promoter-operator
N	any nucleoside
nt	nucleotide(s)
ori	origin of replication

P	promotor
p	plasmid
PAA	polyacrylamide
PAGE	polyacrylamide-gel electrophoresis
Pu	purine
Py	pyrimidine
r	(superscript) resistence
RBS	ribosome binding site
RP	reversed phase
s	(supercript) sensitivity
SD	Shine Dalgarno-sequence
ss-DNA	single-stranded DNA
stp	stop codon
t	terminator
tlc	thin layer chromatography
Tc	tetracycline
TMV	tobacco mosaic virus
wt	wild type

9 References

1. Watson, J. D., Crick, F. H. C.: Nature *171*, 737 (1953)
2. Grunstein, M., Hogness, D. S.: Proc. Natl. Acad. Sci. USA *72*, 3961 (1975)
3. Benton, W. D., Davis, R. W.: Science *196*, 180 (1977)
4. Wallace, R. B. et al.: Nucl. Acids Res. *6*, 3543 (1979)
5. Dodgson, J. B., Wells, R. D.: Biochemistry *16*, 2367 (1977)
6. Szostak, J. W. et al.: Methods Enzymol. *68*, 419 (1979)
7. Ikuta, S. et al.: Nucl. Acids Res. *15*, 797 (1987)
8. Suggs, S. V. et al.: Proc. Natl. Acad. Sci. USA *78*, 6613 (1981)
9. Zon, G. et al.: Nucleic Acids Res. *13*, 8181 (1985)
10. Whitehead, A. S. et al.: Proc. Natl. Acad. Sci. USA *80*, 5387 (1983)
11. Jage, M. et al.: Nucl. Acids Res. *11*, 2325 (1983)
12. Noges, B. E. et al.: Proc. Natl. Acad. Sci. USA *76*, 1770 (1979)
13. Lewis, R. M.: Nucleic Acids Res. *14*, 567 (1986)
14. Anderson, S. et al.: Science *230*, 144 (1985)
15. Ikemura, T.: J. Mol. Biol. *151*, 389 (1981)
16. Takahashi, Y. et al.: Proc. Natl. Acad. Sci. USA *82*, 1931 (1985)
17. Kawase, Y. et al.: Nucleic Acids Res. *14*, 7727 (1986)
18. Saito, S. et al.: Biochem. Biophys. Res. Commun. *141*, 1099 (1986)
19. Sanger, F., Nicklen, S., Coulson, A. R.: Proc. Natl. Acad. Sci. USA *74*, 5462 (1977)
20. Winter, G., Fields, S., Gait, M. J.: Nucl. Acids Res. *9*, 237 (1981)
21. Feinberg, A. P., Vogelstein, B.: Anal. Biochem. *132*, 6 (1983)
22. Goelet, P., Karn, J.: Gene, *29*, 331 (1984)
23. Hutchison, C. A. et al.: J. Biol. Chem. *253*, 6551 (1978)
24. Southern, E. M.: J. Mol. Biol. *98*, 503 (1975)
25. Pauling, L. et al.: Science *25*, 543 (1949)
26. Piratsu, M. et al.: N. Engl. J. Med. *309*, 284 (1983)
27. Kidd, V. J. et al.: Nature *304*, 230 (1983)
28. Trevors, J. T.: Trends in Biotechnology 3, 291 (1985)

29. Kössel, H., Seliger, H.: Fortschr. Chem. Org. Naturst. *32*, 297 (1975)
30. Amarnath, V., Broom, A. D.: Chem. Rev. *77*, 183 (1977)
31. Reese, C. B.: Tetrahedron *34*, 3143 (1978)
32. Narang, S. A.: Tetrahedron *39*, 3 (1983)
33. Pfleiderer, W.: Merck Kontakte *2*, 3 (1983)
34. Ohtsuka, E., Ikehara, M., Söll, D.: Nucl. Acids Res. *10*, 6553 (1982)
35. Smith, J.: American Biotechnology Laboratory *15*, (1983) December
36. Davies, J. E., Gassen, H. G.: Angew. Chem. Int. Ed. Engl. *22*, 13 (1983)
37. Itakura, K., Rossi, J. J., Wallace, R. B.: Ann. Rev. Biochem. *53*, 323 (1984)
38. Sonveaux, E.: Bioorganic Chemistry *14*, 274 (1986)
39. Gassen, H. G., Lang, A.: Chemical and Enzymatic Synthesis of gene fragments, Weinheim, Verlag Chemie 1982
40. Gait, M. J.: Oligonucleotide Synthesis, a practical approach, Oxford, IRL Press 1984
41. Mizuno, Y.: The organic chemistry of nucleic acids. (Studies in organic chemistry, *24*) Tokyo, Kodansha 1986
42. Michelson, A. M., Todd, A. R.: J. Chem. Soc. 2632 (1955)
43. Khorana, H. G.: Bioorg. Chem. *1*, 351 (1978)
44. Khorana, H. G.: Science *203*, 614 (1979)
45. Letsinger, R. L., Mahadevan, V.: J. Am. Chem. Soc. *88*, 5319 (1966)
46. Eckstein, F., Scheit, K. H.: Angew. Chem. Int. Ed. Engl. *6*, 362 (1967)
47. Eckstein, F., Rizk, I.: Chem. Ber. *102*, 2362 (1969)
48. Reese, C. B., Saffhill, R.: J. Chem. Soc. Chem. Commun. 767 (1968)
49. Catlin, J. C., Cramer, F.: J. Org. Chem. *38*, 245 (1973)
50. Itakura, K. et al.: Canad. J. Chem. *51*, 3649 (1973)
51. Itakura, K. et al.: J. Am. Chem. Soc. *97*, 7327 (1975)
52. Miyoshi, K., Itakura, K.: Tetrahedron Lett. *20*, 3635 (1979)
53. Efimov, V. A. et al.: Nucleic Acids Res. *11*, 8369 (1983)
54. Froehler, B. C., Matteucci, M. D.: J. Amer. Chem. Soc. *107*, 278 (1985)
55. Letsinger, R. L. et al. J. Am. Chem. Soc. *97*, 3278 (1978)
56. Beaucage, S. L., Caruthers, M. H.: Tetrahedron Lett. *22*, 1859 (1981)
57. Garegg, P. J. et al.: Chemica Scripta *28*, 280 (1985)
58. Froehler, B. C., Ng, P. G., Mateucci, M. D.: Nucl. Acids Res. *14*, 5399 (1986)
59. Garegg, P. J. et al.: Tetrahedron Lett. *27*, 4051 (1986)
60. Hall, R. H., Todd, A., Webb, R. F.: J. Chem. Soc. 3291 (1957)
61. Frank, R. et al.: Nucleic Acids Res. *11*, 4365 (1983)
62. Ott, J., Eckstein, F.: Nucleic Acids Res. *12*, 9137 (1984)
63. Matthes, H. W. D., Zeneke, W. M., Grundstroem, T.: EMBO J. *3*, 801 (1984)
64. Bannwarth, W., Iaiza, P.: DNA *5*, 413 (1986)
65. Ito, H. et al.: Nucleic Acids Res. *10*, 1755 (1982)
66. Matteucci, M. D., Caruthers, M. H.: Tetrahedron Lett. *21*, 719 (1980)
67. Matteucci, M. D., Caruthers, M. H.: J. Amer. Chem. Soc. *103*, 3185 (1981)
68. Gough, G. R., Brunden, M. J., Gilham, P. T.: Tetrahedron Lett. *22*, 4177 (1981)
69. Adams, S. P. et al.: J. Am. Chem. Soc. *105*, 661 (1983)
70. Crea, R., Horn, T.: Nucleic Acids Res. *8*, 2331 (1981)
71. Miyoshi, K.: Nucleic Acids Res. *8*, 5473 (1980)
72. Potapov, V. K. et al.: Nucleic Acids Res. *6*, 2041 (1979)
73. Greene, T. W.: Protective groups in Organic Synthesis, p. 4, J. Wiley & Sons, New York 1981
74. Watanabe, K. A., Fox, J. J.: Angew. Chem. Int. Ed. *5*, 579 (1966)
75. Khorana, H. G., Turner, A. F., Vizsolyi, J. P.: J. Amer. Chem. Soc. *83*, 686 (1961)
76. Jordan, F., Niv, H.: Nucleic Acids Res. *4*, 697 (1977)
77. Schaller, H. et al.: J. Amer. Chem. Soc. *85*, 3821 (1963)
78. Ti, G. S., Gaffney, B. L., Jones, R. A.: J. Amer. Chem. Soc. *104*, 1316 (1982)
79. Froehler, B. C., Matteucci, M. D.: Nucleic Acids Res. *11*, 8031 (1983)
80. McBride, L. J., Caruthers, M. H.: Tetrahedron Lett. *24*, 2953 (1983)
81. Kume, A. et al.: Nucleic Acids Res. *12*, 8525 (1984)
82. Pless, R. C., Letsinger, R. L.: Nucleic Acids Res. *2*, 773 (1975)
83. Barone, A. D., Tang, J. Y., Caruthers, M. H.: Nucleic Acids Res. *12*, 4051 (1984)

84. Reese, C. B., Ubasawa, A.: Tetrahedron Lett. *21*, 2265 (1980)
85. Reese, C. B., Skone, P. A.: J. Chem. Soc. Perk. Trans. I, 1263 (1984)
86. Himmelsbach, F. et al.: Tetrahedron Lett. *40*, 59 (1984)
87. Engels, J., Mag, M.: Nucleosides Nucleotides *6*, 473 (1987)
88. Büchi, H., Khorana, H. G.: J. Mol. Biol. *75*, 251 (1972)
89. Gaffney, B. L., Jones, R. A.: Tetrahedron Lett. *23*, 2257 (1982)
90. Trichtinger, T., Charubala, R., Pfleiderer, W.: Tetrahedron Lett. *24*, 711 (1983)
91. Sekine, M., Matsuzaki, J., Hata, T.: Tetrahedron Lett. *23*, 5287 (1982)
92. Pon, R. T., Damha, M. J., Ogilvie, K. K.: Tetrahedron Lett. *26*, 2525 (1985)
93. Chattopadhyaya, J. B., Reese, C. B.: J. Chem. Soc. Chem. Commun. 639 (1978)
94. Smith, M. et al.: J. Amer. Chem. Soc. *84*, 430 (1962)
95. Engels, J., Krahmer, U.: Angew. Chem. Int. Ed. Engl. *18*, 942 (1979)
96. Letsinger, R. L. et al.: Tetrahedron *40*, 137 (1984)
97. Duckworth, M. L.: Nucleic Acids Res. *9*, 1691 (1981)
98. Matteucci, M. D., Caruthers, M. H.: J. Amer. Chem. Soc. *103*, 3185 (1981)
99. Kohli, V., Blöcker, H., Köster, H.: Tetrahedron Lett. *21*, 2683 (1980)
100. Barany, G., Merrifield, R. B.: J. Amer. Chem. Soc. *99*, 7363 (1977)
101. van Boom, J. H. et al.: J. Chem. Soc. Chem. Commun. 869 (1971)
102. Itakura, K. et al.: Can. J. Chem. *51*, 3469 (1973)
103. Reese, C. B., Titmas, R. C., Yau, L.: Tetrahedron Lett. *19*, 2727 (1978)
104. Itakura, K., Katagiri, N., Narang, S. A.: Canad. J. Chem. *52*, 3689 (1974)
105. Claesen, C. A. A., Segers, R. P. A. M., Tesser, G. I.: Rec. Trav. Chim. Pays-Bas *104*, 209 (1985)
106. Agarwal, K. L., Riftina, F.: Nucl. Acids Res. *5*, 2809 (1978)
107. van der Marel, G. A. et al.: Tetrahedron Lett. *22*, 3887 (1981)
108. Marugg, J. E. et al.: Nucl. Acids Res. *12*, 8639 (1984)
109. Stawinski, S., Hozumi, T., Narang, S. A.: Canad. J. Chem. *54*, 670 (1976)
110. van Boom, J. H. et al.: Nucleic Acids. Res. *4*, 1047 (1977)
111. Engels, J. et al.: Liebigs Ann. Chem. 745 (1982)
112. Efimov, V. A., Chakhmakhcheva, O. G., Ovchinnikov, Y. A.: Nucleic Acids Res. *13*, 3651 (1985)
113. Efimov, V. A. et al.: Nucleic Acids Res. *14*, 6525 (1986)
114. Daub, G. W., van Tamelen, E. E.: J. Am. Chem. Soc. *99*, 3526 (1977)
115. Caruthers, M. H. et al.: Genetic Engineering, Vol. *4*. 1. Plenum Publishing Comp. (1982)
116. Sinha, N. D., Biernat, J., Köster, H.: Tetrahedron Lett. *24*, 5843 (1983)
117. Kuyl-Yeheskiely, E. et al.: Recl. Trav. Chim. Pays-Bas *105*, 505 (1986)
118. Fourrey, J. L., Varenne, J.: Tetrahedron Lett. *26*, 1217 (1985)
119. Uhlmann, E., Engels, J.: Tetrahedron Lett. *27*, 1023 (1986)
120. Uhlmann, E., Engels, J.: Chemica Scripta *26*, 217 (1986)
121. Horn, T., Urdea, M. S.: Tetrahedron Lett. *27*, 4705 (1986)
122. Tanaka, T., Yamada, Y., Ikehara, M.: Tetrahedron Lett. *27*, 3267 (1986)
123. Connolly, B. A.: Tetrahedron Lett. *28*, 463 (1987)
124. Agrawal, S., Christodoulou, C., Gait, M. J.: Nucleic Acids Res. *14*, 6227 (1986)
125. Reddy, M. P., Rampal, J. B., Beaucage, S. L.: Tetrahedron Lett. *28*, 23 (1987)
126. van Boom, J. H. et al.: J. Chem. Soc. Chem. Commun. 167 (1976)
127. Uhlmann, E., Pfleiderer, W.: Tetrahedron Lett. *21*, 1181 (1980)
128. Uhlmann, E., Pfleiderer, W.: Helv. Chim. Acta *64*, 1688 (1981)
129. Schulhof, J. C., Molko, D., Teoule, R.: Nucl. Acids Res. *15*, 397 (1986)
130. Schulhof, J. C., Molko, D., Teoule, D.: Tetrahedron Lett. *28*, 51 (1987)
131. Reese, C. B., Zard, L.: Nucleic Acids Res. *9*, 4611 (1981)
132. Atkinson, T., Smith, M.: in Oligonucleotide Synthesis, a practical approach (ed. Gait, M.), p. 35 IRL press, Oxford 1984
133. Sinha, N. D. et al.: Nucleic Acids Res. *12*, 4539 (1984)
134. Sproat, B. S., Gait, M. J.: in Oligonucleotide Synthesis, a practical approach (ed. Gait, M.), p. 83 IRL press, Oxford 1984
135. Sanchez-Pescador, R., Urdea, M. S.: DNA *3*, 339 (1984)
136. Maniatis, T., Fritsch, E. F., Sambrook, J.: J. Molecular cloning. A Laboratory Manual, Cold Spring Harbor Laboratory Press, New York 1982

137. Gait, M. J. et al.: in Chemical and Enzymatic synthesis of Gene Fragments, A Laboratory Manual (eds. Gassen, H. G., Lang, A.), p. 37 Verlag Chemie, Weinheim 1982
138. Wu, R. et al.: in Oligonucleotide Synthesis, a practical approach (ed. Gait, M.) IRL press, Oxford 1984
139. Frank, R., Köster, H.: Nucleic Acids Res. *6*, 2069 (1979)
140. Jing, G. Z., Liu, A. P., Leung, W. C.: Anal. Biochem. *155*, 376 (1986)
141. Farretti, L. et al.: Proc. Natl. Acad. Sci. USA *83*, 599 (1986)
142. Brown, P. R.: High Pressure Liquid Chromatography, Academic Press, New York 1973
143. McLaughlin, L. W., Krusche, J. U.: in Chemical and Enzymatic synthesis of Gene Fragments: A Laboratory Manual (eds. Gassen, H. G.)
144. McLaughlin, L. W., Piel, N.: in Oligonucleotide Synthesis, a practical approach (ed. Gait, M.) p. 117, IRL press, Oxford 1984
145. Schott, H.: Nucleobases, Nucleosides, Nucleotides in High Performance Liquid Chromatography in Biochemistry (eds. Henschen, A., Hupe, K. P., Lottspeich, F., Voelter, W.) p. 414, Verlag Chemie, Weinheim 1985
146. Zon, G., Thompson, J. A.: Bio Chromatography *1*, 22 (1986)
147. Fritz, H. J. et al.: Biochem. *17*, 1257 (1978)
148. Haupt, W., Pingoud, A.: J. Chromatogr. *260*, 419 (1983)
149. Görtz, H. H., Seliger, H.: Angew. Chem. Int. Ed. Engl. *20*, 681 (1981)
150. Newton, C. R. et al.: Anal. Biochem. *129*, 22 (1983)
151. Agarwal, K. L. et al.: Nature *227*, 27 (1970)
152. Khorana, H. G. et al.: J. Biol. Chem. *251*, 565 (1976)
153. van de Sande, H. J. et al.: J. Biol. Chem. *251*, 571 (1976)
154. Minamoto, K. et al.: J. Biol. Chem. *251*, 587 (1976)
155. Agarwal, K. L. et al.: J. Biol. Chem. *251*, 599 (1976)
156. Jay, E. et al.: J. Biol. Chem. *251*, 609 (1976)
157. Agarwal, K. L. et al.: J. Biol. Chem. *251*, 624 (1976)
158. Sekiya, T. et al.: J. Biol. Chem. *251*, 634 (1976)
159. Loewen, P. C. et al.: J. Biol. Chem. *251*, 642 (1976)
160. Panet, A. et al.: J. Biol. Chem. *251*, 651 (1976)
161. Caruthers, M. H. et al.: J. Biol. Chem. *251*, 658 (1976)
162. Kleppe, R. et al.: J. Biol. Chem. *251*, 667 (1976)
163. Ramamoorthy, B. et al.: J. Biol. Chem. *251*, 676 (1976)
164. Brown, E. L. et al.: Chemical Synthesis and Cloning of a Tyrosine tRNA Gene, in Methods in Enzymology (ed. Wu, R.) *68*, 109 (1979)
165. Köster, H. et al.: Hoppe-Seylers Z. Physiol. Chem. *356*, 1585 (1975)
166. Itakura, K. et al.: Science *198*, 1056 (1977)
167. Crea, R. et al.: Proc. Natl. Acad. Sci. USA *75*, 5765 (1978)
168. Goeddel, D. V. et al.: Proc. Natl. Acad. Sci. USA *76*, 106 (1979)
169. Edge, M. D. et al.: Nature *292*, 756 (1981)
170. Ikehara, M. et al.: Proc. Natl. Acad. Sci. USA *81*, 5956 (1984)
171. Engels, J. W. et al.: Pure & Appl. Chem. *59*, 437 (1987)
172. Wetzel, R. et al.: Gene *16*, 63 (1981)
173. Sung, W. L. et al.: Biochem. Cell Biol. *64*, 133 (1986)
174. Engels, J. W. et al.: Nucleosides Nucleotides *6*, 185 (1987)
175. Urdea, M. S. et al.: Proc. Natl. Acad. Sci. USA *80*, 7461 (1983)
176. Vlasuk, G. P. et al.: J. Biol. Chem. *261*, 4789 (1986)
177. Smith, J. et al.: Nucleic Acids Res. *10*, 4467 (1982)
178. Ivanov, I., Gigova, L., Jay, E.: FEBS Letters *210*, 56 (1987)
179. Oka, T. et al.: Proc. Natl. Acad. Sci. USA *82*, 7212 (1985)
180. Kempe, T. et al.: Bio/Technology *4*, 565 (1986)
181. Kishimoto, F. et al.: Gene *45*, 311 (1986)
182. Ovchinnikov, Y. A. et al.: Gene *31*, 65 (1984)
183. Brousseau, R. et al.: Gene *17*, 279 (1982)
184. Miyoshi, K. et al.: Nucleic Acids Res. Symp. Ser. *10*, 197 (1981)
185. Cravador, A. et al.: Biochemie *67*, 829 (1985)

186. Alton, K. et al.: in The Biology of the Interferon System (eds. De Maeyer, E., Schellekens, H.) Elsevier Science Publishers B. V., p. 119 (1983)
187. Engels, J., Leineweber, M., Uhlmann, E.: Angew. Chem. Int. Ed. Engl. *23*, 961 (1984)
188. Jay, E. et al.: J. Biol. Chem. *259*, 6311 (1984)
189. Tanaka, S. et al.: Nucleic Acids Res. *11*, 1707 (1983)
190. Gray, P. W. et al.: Nature, *312*, 721 (1984)
191. Edge, M. D. et al.: Nucleic Acids Res. *11*, 6419 (1983)
192. Zurawski, S. M. et al.: Gene *49*, 61 (1986)
193. Porter, A. G. et al.: DNA *5*, 137 (1986)
194. Camble, R., et al.: Biochem. Biophys. Res. Commun. *134*, 1404 (1986)
195. Leung, W. C. et al.: in Biol. Interferon Syst. Proc TNO-ISIR Meet., (eds. Kirchner et al.) Elsevier Science Publishers B. V., p. 419 (1985)
196. Nagase, Y. et al.: Nucleic Acids Symp. Ser. *12*, 83 (1983)
197. Wetzel, R. et al.: Biochem. *19*, 6096 (1980)
198. Mandecki, W. et al.: Proc. Natl. Acad. Sci. USA *82*, 3543 (1985)
199. Mullenbach, G. T. et al.: J. Biol. Chem. *261*, 719 (1986)
200. Sung, W. L. et al.: Nucleic Acids Res. *14*, 6159 (1986)
201. Sproat, B. S., Gait, M. J.: Nucleic Acids Res. *13*, 2959 (1985)
202. Emblad, A. et al.: In "Third European Congress on Biotechnology", Vol. III, p. 287, Verlag Chemie, Weinheim (1984)
203. Buell, G. et al.: Nucleic Acids Res. *13*, 1923 (1985)
204. Allen, G., Paynter, C. A., Winther, M. D.: J. Cell Sci., Suppl. *3*, 29 (1985)
205. Peters, M. A. et al.: Gene *35*, 83 (1985)
206. Nambiar, K. P. et al.: Science *223*, 1299 (1984)
207. Jigami, Y. et al.: Gene *43*, 273 (1986)
208. Bülow, L., Mosbach, K.: FEBS Lett. *210*, 147 (1987)
209. Ikehara, M. et al.: Proc. Natl. Acad. Sci. *83*, 4695 (1986)
210. Gearing, D. P., McMullen, G., Nagley, P.: Biochem. Int. *10*, 907 (1985)
211. Kempe, T. et al.: Gene *39*, 239 (1985)
212. Hostomsky, Z., Smrt, J., Paces, V.: Gene *39*, 269 (1985)
213. Tang, J. et al.: Sci. Sin. Ser. B (Engl. Ed.) *27*, 466 (1984)
214. Tanaka, S. et al.: Nucl. Acids Res. *10*, 1741 (1982)
215. Dobrynin, V. N. et al.: Bioorg. Khim. *7*, 1745 (1981)
216. Brodin, P. et al.: Biochem. *25*, 5371 (1986)
217. Roberts, D. M. et al.: Biochem. *24*, 5090 (1985)
218. Theriault, N. Y., Tomich, C.-S., Wierenga, W.: Nucleosides Nucleotides *5*, 15 (1986)
219. Bergmann, C. et al.: Biol. Chem. Hoppe-Seyler *367*, 731 (1986)
220. Rink, H. et al.: Nucl. Acids Research *12*, 6369 (1984)
221. Fortkamp, E. et al.: DNA *5*, 511 (1986)
222. Doel, M. T. et al.: Nucleic Acids Res. *8*, 4575 (1980)
223. Kumarev, V. P. et al.: FEBS Lett. *114*, 273 (1980)
224. Ohlrogge, J. B. et al.: Biochem. Soc. Trans. *14*, 579 (1986)
225. Miura, K. et al.: Jpn. J. Cancer Res. *77*, 45 (1986)
226. Altmann, S., Smith, J. D.: Nature New Biol. *233*, 35 (1971)
227. Ryan, M. J. et al.: J. Biol. Chem. *254*, 5817 (1979)
228. Engels, J. W. et al.: Protein Engineering, *1*, 195 (1987)
229. Fiers, W. et al.: Nature *260*, 500 (1976)
230. Bertrand, K. et al.: Science *189*, 22 (1975)
231. Mantei, N. et al.: Gene *10*, 1 (1980)
232. De Maeyer, E. et al.: Proc. Natl. Acad. Sci. USA *79*, 4256 (1982)
233. Caruthers, M. H.: Contemp. Top. Polym. Sci. *5*, 55 (1984)
234. Stewart, A. G. et al.: Nucleic Acids Res. *11*, 6597 (1983)
235. Rossi, J. J. et al.: J. Biol. Chem. *257*, 9226 (1982)
236. Khorana, H. G.: Pure Appl. Chem. *17*, 349 (1968)
237. Scarpulla, R. C., Narang, S., Wu, R.: Anal. Biochem. *121*, 356 (1982)
238. Chakhmakhcheva, O. G. et al.: Chemica Scripta. *26*, 31 (1986)
239. Rossi, J. J. et al.: Proc. Natl. Acad. Sci. USA *80*, 3203 (1983)

240. Nagai, K., Thoegersen, H. C.: Nature *309*, 810 (1984)
241. Connolly, B. A. et al.: Biochem. *23*, 3443 (1984)
242. Efcavitch, J. W., McBride, L. J., Eadie, J. S.: in Biophosphates and their analogues — Synthesis, structure, metabolism and activity (eds. Bruzik, K. S., Stec, W. J.), p. 65, Elsevier Science Publishers, Amsterdam 1987
243. Hein, F., Jansen, H. W., Uhlmann, E.: Nucleosides Nucleotides, *6*, 489 (1987)
244. Chakhmakhcheva, O. G., Efimov, V. A., Ovchinnikov, Y. A.: Nucleosides and Nucleotides, *6*, 321 (1987)
245. Oliphant, A. R., Nussbaum, A. L., Struhl, K.: Gene *44*, 177 (1986)
246. Derbyshire, K. M., Salvo, J. J., Grindley, N. D. F.: Gene *46*, 145 (1986)
247. Zoller, M. J., Smith, M.: Nucl. Acids Res. *10*, 6487 (1982)
248. Fritz, H. J.: "New developments in mutation construction and chemical gene synthesis" presented at the 17th FEBS meeting Berlin (West) *367*, 86 (1986)
249. Kramer, W. et al.: Nucl. Acids Res. *12*, 9441 (1984)
250. Snopek, T. J. et al.: Biochem. Biophys. Res. Commun. *68*, 417 (1976)
251. Brennan, C. A., Monthey, A. E., Gumport, R. I.: Methods in Enzymology *100*, 38 (1983)
252. Brennan, C. A., Gumport, R. I.: Nucleic Acids Res. *13*, 8665 (1985)
253. Tessier, D. C., Brousseau, R., Vernet, T.: Analyt. Biochem. *158*, 171 (1986)
254. Cosstick, R., McLaughlin, L. W., Eckstein, F.: Nucleic Acids Res. *12*, 1791 (1984)
255. Zeh, D., Seliger, H.: Nucleosides Nucleotides, *6*, 485 (1987)
256. Grundstroem, T. et al.: Nucleic Acids Res. *13*, 3305 (1985)
257. Narang, S. A. et al.: Biochem. Biophys. Res. Commun. *134*, 407 (1986)
258. Smith, M.: Ann. Rev. Genet. *19*, 423 (1985)
259. Kramer, B., Kramer, W., Fritz, H. J.: Cell *38*, 879 (1984)
260. Zoller, M. J., Smith, M.: DNA *3*, 479 (1984)
261. Kramer, W., Schughart, K., Fritz, H. J.: Nucleic Acids Res. *10*, 6475 (1982)
262. Carter, P., Bedouelle, H., Winter, G.: Nucleic Acids Res. *13*, 4431 (1985)
263. Kunkel, T. A.: Proc. Natl. Acad. Sci. USA *82*, 488 (1985)
264. Taylor, J. W., Ott, J., Eckstein, F.: Nucleic Acids Res. *13*, 8764 (1985)
265. Matteucci, M. D., Heynecker, H. L.: Nucleic Acids Res. *11*, 3113 (1983)
266. Wells, J. A., Vasser, M., Powers, D. B.: Gene *34*, 315 (1985)
267. Cline, S. W., Yarus, M., Wier, P.: DNA *5*, 37 (1986)
268. Grosjean, H., Fiers, W.: Gene *18*, 199 (1982)
269. Ikemura, T.: J. Mol. Biol. *158*, 573 (1982)
270. Maruyama, T. et al.: Nucl. Acids Res. *14* (Supplement), r 151 (1986)
271. Robinson, M. et al.: Nucl. Acids Res. *12*, 6663 (1984)
272. Tokunaga, T. et al.: Gene *39*, 117 (1985)
273. Coulombe, B., Skup, D.: Gene *46*, 89 (1986)
274. Kniskern, P. J. et al.: Gene *46*, 135 (1986)
275. Sharp, P. M., Tuohy, T. M., Mosurski, K. R.: Nucleic Acids Res. *14*, 5125 (1986)
276. O'Neill, G. et al.: Gene *44*, 325 (1986)
277. Vieira, J., Messing, J.: Gene *19*, 259 (1982)
278. Messing, J. et al.: Proc. Natl. Acad. Sci. USA *74*, 3642 (1977)
279. Roberts, R. J.: Nucleic Acids Res. Supplement *13*, r 165 (1985); *15*, Suppl. r 189 (1987)
280. Kessler, C., Höltke, H. J.: Gene *47*, 1 (1986)
281. Jay, E. et al.: Proc. Natl. Acad. Sci. *81*, 2290 (1984)
282. Hopp, T. P., Woods, K. R.: Proc. Natl. Acad. Sci. USA *78*, 3824 (1981)
283. Chou, P. Y., Fasman, G. D.: Annu. Rev. Biochem. *47*, 251 (1978)
284. Szorka, P. R. et al.: DNA *5*, 11 (1986)
285. Landon, M.: in Methods in Enzymology, Vol. *47*, (eds. Grossman, L., Moldava, K.) Academic Press, New York, p. 145–149
286. Lazure, C. et al.: Can. J. Biochem. Cell. Biol. *61*, 501 (1983)
287. Guo, K. H. et al.: Gene *29*, 251 (1984)
288. Sung, W. L. et al.: Proc. Natl. Acad. Sci. USA *83*, 561 (1986)
289. Abrahamsen, L. et al.: Nucleic Acids Res. *14*, 7487 (1986)
290. Dodt, J. et al.: FEBS Lett. *202*, 373 (1986)
291. Gray, G. L. et al.: Gene *39*, 247 (1985)

292. Blobel, G., Dobberstein, B.: J. Cell. Biol. *67*, 835 (1975)
293. Uhlmann, E., Engels, J., Wetekam, W.: Nucleosides Nucleotides *4*, 259 (1985)
294. Nilsson, B. et al.: Nucleic Acids Res. *13*, 1151 (1985)
295. Ohmura, K. et al.: Nucleic Acids Res. *12*, 5307 (1984)
296. Zsebo, K. M. et al.: J. Biol. Chem. *261*, 5858 (1986)
297. Emr, S. D. et al.: Proc. Natl. Acad. Sci. USA *80*, 7080 (1983)
298. Perlman, D., Halvorson, H. O.: J. Mol. Biol. *167*, 391 (1983)
299. von Heijne, G.: J. Mol. Biol. *173*, 243 (1984)
300. Lee, N. et al.: Nucleic Acids Res. *12*, 6797 (1984)
301. Shen, S.-H.: Proc. Natl. Acad. Sci. USA *81*, 4627 (1984)
302. Thim, L. et al.: Proc. Natl. Acad. Sci. USA *83*, 6766 (1986)
303. Bradbury, A. F., Finnie, M. D. A., Smyth, D. G.: Nature *298*, 686 (1982)
304. Hawley, D. K., McClure, W. R.: Nucleic Acids Res. *11*, 2237 (1983)
305. Engels, J., Leineweber, M., Uhlmann, E.: in Organic Synthesis: an interdisciplinary challenge (eds. Streith, J., Prinzbach, H., Schill, G.) Blackwell Scientific Publications, p. 205 (1985)
306. Fritz, H. J.: Nucleic Acids Res. Spec. Publ. *4*, s243 (1978)
307. Dobrynin, V. N. et al.: Nucleic Acids Res. Symp. Ser. *7*, 365 (1980)
308. deHaseth, P. L. et al.: Nucleic Acids Res. *11*, 773 (1983)
309. Rommens, J. et al.: Nucleic Acids Res. *11*, 5921 (1983)
310. Windass, J. D. et al.: Nucleic Acids Res. *10*, 6639 (1982)
311. Deuschle, U. et al.: The EMBO J. 5, 2987 (1986)
312. Jacob, F., Monod, J.: J. Mol. Biol. *3*, 318 (1961)
313. Sadler, J. R., Sasmor, H., Betz, J. L.: Proc. Natl. Acad. Sci. *80*, 6785 (1983)
314. Simons, A. et al.: Proc. Natl. Acad. Sci. USA *81*, 1624 (1984)
315. Leineweber, M., Engels, J., Uhlmann, E.: in preparation
316. Brosius, J., Holy, A.: Proc. Natl. Acad. Sci. USA *81*, 6929 (1984)
317. Besse, M., v. Wilcken-Bergmann, B., Müller-Hill, B.: The EMBO J. *5*, 1377, (1986)
318. Simons, G. et al.: Gene *28*, 55 (1984)
319. Flock, J. I. et al.: Mol. Gen. Genet. *195*, 246 (1984)
320. Shine, J., Dalgarno, L.: Nature *254*, 34 (1975)
321. Jay, G. et al.: Proc. Natl. Acad. Sci. USA *78*, 5543 (1981)
322. Jay, E. et al.: Nucleic Acids Res. *10*, 6319 (1982)
323. Dworkin-Rastl, E., Swetly, P., Dworkin, M. B.: Gene *21*, 237 (1983)
324. Stanssens, P., Remaut, E., Fiers, W.: Gene *36*, 211 (1985)
325. Marquis, D. M., Smolec, J. M., Katz, D. H.: Gene *42*, 175 (1986)
326. Chambers, I. et al.: The EMBO J. *5*, 1221 (1986)
327. Schottel, J. L., Sninsky, J. J., Cohen, S. N.: Gene *28*, 177 (1984)
328. Looman, A. C. et al.: Nucleic Acids Res. *14*, 5481 (1986)
329. Tessier, L.-H. et al.: Nucleic Acids Res. *12*, 7663 (1984)
330. Zucker, M., Stiegler, P.: Nucleic Acids Res. *9*, 133 (1981)
331. Nussinov, R. et al.: SIAM J. Appl. Math. *35*, 68 (1978)
332. Papanicolaou, C., Gouy, M., Ninio, J.: Nucleic Acids Res. *12*, 31 (1984)
333. Nucleic Acids Res. Vol. 10, 1–456 (1982); Vol. 12, 1–853 (1984); Vol. 14, 1–621 (1986)
334. Young, R.: J. Biol. Chem. *254*, 12725 (1979)
335. Jones, S. S. et al.: Bio/Technology *5*, 67 (1987)
336. Mutter, M.: Angew. Chem. Int. Ed. *97*, 639 (1985)

Strain Improvement in Industrial Microorganisms by Recombinant DNA Techniques

Helmut Schwab
Institut für Biotechnologie, Mikrobiologie und Abfalltechnologie
Technische Universität Graz, A-8010 Graz, Austria

The development of highly productive microbial strains is a prerequisite for efficient biotechnological processes. Up to the present time, strain improvement was mainly based on induced mutagenesis and only during the last few years has genetic recombination based on protoplast fusion become an additional practicable strategy.

In general, recombinant DNA techniques provide an enormous potential for strain improvement in industrial microorganisms. Basic cloning systems have now been developed for various microbial strains used in industrial processes. Not only bacteria and yeasts, but also many biotechnologically important species of filamentous fungi are now susceptible to rDNA techniques.

rDNA technology has already been successfully applied to the biotechnological production of polypeptides used pharmaceutically and originating from higher eukaryotes. For such "high price — low volume" products, well-developed host systems such as *E. coli* are usually employed. The scale of production is usually small. Thus these processes are more or less comparable with laboratory operations. However, special problems have to be taken into account when genetically engineered strains are used in large-scale processes.

This review presents an overview of the current developments of cloning systems in various groups of microorganisms. Furthermore, general problems pertaining to the industrial use of recombinant microorganisms and current strategies for strain improvement by rDNA techniques are discussed.

1 Introduction

For thousands of years the metabolic capabilities of microorganisms have been used to produce food and beverages. However, it was one of the revolutions of the 20th century that the technical use of living organisms evolved to a highly sophisticated technology — biotechnology.

A few years ago the European Federation of Biotechnology defined biotechnology as the "integrated use of biochemistry, microbiology and engineering sciences in order to achieve technological (industrial) applications of the capabilities of microorganisms, cultured tissue cells and parts thereof" [1]. Although there is now a clear trend to extend the field of biotechnology to the general use of any living organism (or parts), microorganisms still hold a dominant position in industrial and environmental applications.

The most interesting feature of microorganisms is their capability of carrying out a wide variety of metabolic activities which provide the basis for the potential to synthesize a broad spectrum of products of commercial value. However, in its natural environment, the metabolism of a living cell is predominantly focused on optimal survival and proper reproduction. In general, this means that organisms use their biochemical machinery very efficiently and synthesize only those metabolites and cell constituents in the minimum amounts required for maintenance and growth. The requirements of a biotechnological process are, however, contrary to these facts since in most cases one specific compound should be produced in large amounts.

In general, the overproduction of specific compounds can be obtained by impairing the naturally existing fine balance of metabolic activities. This can be achieved principally in two ways:

1) Alteration of the normal growth environment to special, often unnatural conditions.
2) Genetic manipulations.

The adjustment of process variables to provide very specific conditions may cause drastic changes in the metabolism of organisms which might result in a considerable overproduction of particular compounds [2-4]. Nevertheless, the range of alterations of metabolism is limited being strictly connected with the inherent genetic constitution of an organism. It is therefore obvious that in addition to the investigation of suitable conditions, genetic strain improvement programs are in most cases integral parts of successful strategies for the development of modern bioprocesses [5].

The useful benefits of "classical" genetic tools, namely induced mutations and sexual or parasexual recombination have already been proven in many cases [6]. However, the newly developed methods of recombinant DNA technology have opened up new horizons for the construction of highly productive strains.

In this review the problems and perspectives pertaining to the development and the application of recombinant DNA systems to industrial microorganisms which are not usually well characterized genetically will be discussed. A further aspect will be the problem of the use of genetically engineered organisms in large-scale bioprocesses.

2 General Considerations on Strain Improvement

2.1 Mutation — Selection

Genetic manipulation of microorganisms was for a long time mainly characterized by breeding programs based on mutation and selection. Although mutagenesis is a random process with almost no possibility of determining the site of action within the genetic material, the strategy of mutation has been shown in many cases to be extremely valuable for strain improvement. The main benefit of mutation is the blocking of undesirable enzyme activities and the elimination of negatively acting regulations. However, in most cases time consuming and labour intensive "random" screening procedures are required often entailing tremendous efforts, persistence, and skill in the art of microbiology. One typical example of this strategy is the production of penicillin. The productivity of this process between 1950 and today has been increased to an extent of more than two orders of magnitude [7-10].

With increasing biochemical and genetic knowledge of metabolism and its regulation, more "rational" selection methods based on predicted biochemical and genetic behaviour of desirable mutants could be developed making mutation-selection strategies more efficient and less time-, work-, and money-consuming. Although several reports exist on the development of rational screening procedures in the field of secondary metabolite production [11-15], much more detailed knowledge on the formation of primary metabolites is present and thus provides a better basis for rational selection methods. Very typical examples for the successful application of rational screening procedures in industrial practice are found in the field of amino acid production [12,13,16].

The rapid development of biological sciences has also had a great impact on "simple" mutation-selection strategies. Together with the integration of modern developments the fields of automation and miniaturization, as well as with the application of statistics and computer-aided data handling, even random screening procedures today represent quite sophisticated activities [17,18].

2.2 Genetic Recombination

Recombination generally provides an exchange of genetic information between different individuals. The most efficient natural recombination systems are connected with sexual reproduction processes in higher eukaryotes.

Sexual reproduction does not exist in prokaryotes and in the case of many lower eukaryotes, only very inefficient or no actual sexual processes are known. Therefore, genetic recombination in microorganisms is very often limited to parasexual mechanisms. However, parasexual mechanisms usually suffer from low frequencies and are only found in a limited number of organisms. In addition, proper genetic markers for efficient selection of recombinants are frequently not available. Therefore, over a long period of time, breeding of microbial strains by sexual or parasexual recombination has not attained a significant role in industrial processes.

The low frequencies of parasexual processes are usually the result of a low efficiency of the initial step — cell fusion. Therefore, induced fusion of cells via their protoplasts can result in a tremendous increase in the formation of hybrids. Starting with the first reports on successful controlled fusion of fungal and bacterial protoplasts [19–21], protoplast fusion provided a real breakthrough for the application of recombination-based breeding systems for industrial microorganisms [22–30]. In addition to the possibility of achieving high recombination frequencies which even permit work without having genetic markers available for the selection of recombinants, protoplast fusion techniques offer the great advantage of being easily applicable to newly isolated, not fully characterized strains.

As protoplast fusion became a feasible technique, naturally existing barriers to gene transfer could be overcome. Some hopes have thus been placed in this method in order to obtain interspecies hybrids. Several reports exist on successful interspecies protoplast fusion mainly with fungi, but also with bacteria [25, 31–35]. However, such hybrids were in most cases found to be very unstable. If stable recombinants could be obtained, then this was possible only with very closely related strains. As a consequence of this, the main benefit of protoplast fusion for industrial practice has been the possibility of combining desired mutant alleles present in different mutants of a species which were obtained from mutagenesis programs. For example, two strains of *Cephalosporium acremonium* were fused. One strain was a low-titre, rapidly growing, spore forming strain requiring methionine; the other was a high-titre, slow growing asporogenous strain which could use inorganic sulfate. Among the progeny a recombinant could be found which grew rapidly, sporulated, could utilize inorganic sulfate, and produced more cephalosporin C than the high-titre parent [27].

2.3 Recombinant DNA

In general, recombinant DNA technology (genetic engineering) represents the sum of newly developed methods which allow the isolation, characterization, in vitro manipulation, transfer and reintegration of genetic material into living cells, and the expression of specific genes in a controlled manner. Furthermore, the rDNA techniques make it possible to overcome almost any barriers with respect to the transfer of specific genes between organisms. With the exception of some problems with post-translational modifications, the expression of genes in heterologous systems can in principle also be solved.

The enormous potential of genetic engineering has been demonstrated with several startling examples connected with the production of mammalian proteins by microorganisms. Human insulin and interferon preparations are already on the market and a series of further hormones and peptides of pharmaceutical interest including proteins of the immune system, blood proteins, peptides of the neurosystem and also proteins used as vaccines are presently under development [36-41]. However, in most cases *Escherichia coli*, and in a few further cases such well characterized systems as *Bacillus subtilis* and *Saccharomyces cerevisiae* have been used as the host organisms [42-44].

Various theoretical concepts and strategies for applications of rDNA technology to the improvement and development of industrial processes employing microorganisms have been worked out. Although basic cloning systems are now developed for a number of commercial microbial systems, only few reports exist on the use of genetically engineered strains for large scale processes. Some basic problems of the development of cloning systems and of the realization of strain improvement strategies using rDNA technology will be discussed in detail in the following sections.

3 Basic Aspects of rDNA Technology

From a general point of few, all main strategies for the application of rDNA techniques for strain improvement are based on the controlled expression of specific genes. Such genes can be derived either from homologous or from heterologous systems. The general steps of a cloning strategy based on *E. coli* shuttle vectors are shown in Fig. 1.

The first steps involve the preparation of specific DNA fragments which carry the desired genetic information. This is in most cases done by the construction and subsequent screening of genomic or cDNA libraries from selected organisms. In rare cases chemical synthesis of DNA fragments based on protein sequence data will be an alternative strategy.

Standard cloning procedures using a great variety of well established techniques and well developed vectors and *E. coli* host strains are used exclusively for this work. In addition, the characterization of isolated DNA fragments including sequencing, and the subsequent modifications are based on standard techniques using *E. coli* systems. Furthermore, the joining of the DNA fragments to the vector for the target host, and the selection and analysis of proper recombinant molecules will be in most cases also performed with standard *E. coli* systems. Finally, part of the work for development and construction of suitable vectors for the target organisms will be carried out in the same manner. For reviews and basic protocols on such cloning procedures the reader should refer to the references [45-52].

Only the last steps, namely transfer and selection, have to be performed with methods which must be usually specifically developed for the host organism used.

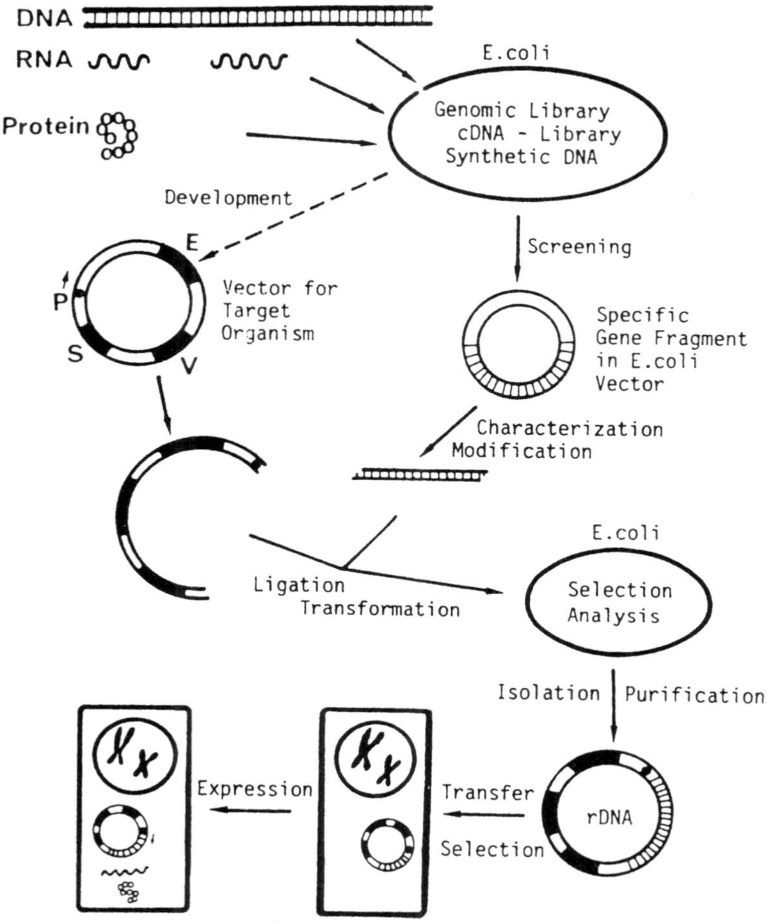

Fig. 1. General steps of rDNA technology. DNA fragments encoding the desired genetic information are usually isolated by screening gene libraries which have been constructed from relevant organisms in *E. coli*. Chemical synthesis of DNA fragments based on protein sequence data is an alternative strategy. The respective DNA fragments are usually characterized and modified using *E. coli* systems and finally fused to a suitable vector molecule which may carry an expression system for the target host. After transfer of the recombinant DNA to the target host and selection of proper clones, the recombinant microorganisms are further tested for their ability to express the newly introduced genetic information. For further explanations see text. The abbreviations used in the drawing of the vector are: E: replication and selection system for *E. coli*, V: system for vegetative maintenance in the target host (replication or integration), S: selection marker for target organism. P: signals directing expression of DNA fragments in the target host (Promoter etc.)

In addition, the expression system has also to be capable of being functional in the target organism.

From this general viewpoint, it can be concluded that for the application of rDNA technology to a specific microbial system three primary prerequisites must be fulfilled:

1) Establishment of sufficiently effective DNA transfer systems.
2) Availability of suitable vectors including selection systems.
3) Availability of functional control elements for gene expression.

4 DNA Transfer Systems

Any strategy for in vitro manipulation of DNA molecules requires a mechanism which allows the subsequent transfer of DNA into living cells of an organism. In general, the following possibilities are available:

1) Transformation.
2) In vitro packaging of DNA into phage heads and transfer to host cells via phage infection.
3) Microinjection.
4) Conjugative transfer from an intermediate host.

Microinjection is only practicable with very large cells. This method has, therefore, no relevance for microbial systems. In addition, the bacteriophage-mediated transfer via in vitro packaging is only well developed for the bacteriophage λ system and is again restricted to special laboratory strains of *E. coli*. For applications with industrial microorganisms, only the remaining two systems of transformation and conjugation are of practical interest.

4.1 Transformation

Transformation was originally found to be a natural mechanism allowing the uptake of free DNA into bacterial cells. This mechanism is usually dependent on a specific physiological state of the cells (competence) and was only found in a limited number of bacterial species [53]. Artificial treatment of cells with certain chemicals, usually mono- or bivalent cations, allowed the establishment of transformation systems for various further bacterial species. For example, transformation of *E. coli* [54] and some other Gram-negative bacteria [55–59] could be obtained after treatment of cells with Ca^{++} or Mg^{++}. However, with the exception of several *E. coli* laboratory strains, only rather low frequencies could usually be obtained, whereas for a substantial number of bacterial species, transformation could not be achieved at all.

A major step forward to the development of more universal transformation systems was attained by the introduction of protoplast techniques. With the exception of the Gram-negative bacteria, protoplast formation and regeneration can be relatively easily realized. Therefore, for almost all Gram-positive bacteria and for fungal systems transformation via protoplasts is now the method of choice for the introduction of DNA. All procedures follow a general 3-step scheme:

1) Formation of protoplasts (spheroplasts) in osmotic stabilized media with suitable cell wall lytic enzymes.
2) Introduction of DNA into protoplasts mediated by polyethyleneglycol (PEG 4,000 — PEG 6,000) in the presence of Ca^{++}.
3) Regeneration of protoplasts and selection of transformants.

General protocols for protoplast transformation of *Bacillus* species, streptomycetes and yeasts are given in references [60–62]. Similar concepts have been developed for filamentous fungi and all methods are in principle analogous to the well developed transformation procedures for *Neurospora crassa* [63] and *Aspergillus nidulans* [64]. A collection of commercially available cell wall lytic enzyme preparations provide the basis for efficient protoplast formation from various fungal strains [65].

Rapid degradation of transforming DNA by extracellular and intracellular nucleases was suspected to be one of the reasons for low transformation frequencies with several microbial systems. Encapsulation of DNA within synthetic lipid vesicles and subsequent fusion with protoplasts has been suggested as a method offering good possibilities of overcoming such problems. Some reports on the successful use of liposomes are present [66-68]. However, technical difficulties in preparing and using liposomes are reasons for low levels of reproducibility which prevent the extensive use of this technology.

An alternative to protoplast transformation in eukaryotic microorganisms has been developed for yeast. Intact cells can be transformed in the presence of PEG after treatment of cells with lithium acetate [69]. In a similar procedure, treatment of germinating conidia of *Neurospora crassa* with lithium acetate permitted the introduction of plasmid DNA into this organism [70].

Transformation mediated by physical methods based on electrical breakdown of cell membranes (electroporation) has been developed for plant and animal cells [71]. Electrically induced DNA transfer into microbial cells has been recently demonstrated but does not seem to be that effective [72]. However, further optimization of electrical and media parameters may result in higher yields and thus could provide good possibilities for this method to become an alternative to chemical transformation methods.

An extensive list of transformation systems in various species of microorganisms was recently given by Imanaka [73].

4.2 Conjugative Plasmid Transfer

As has already been mentioned above, it is not possible to transform many bacteria, particularly Gram-negative strains. Fortunately, the conjugation system of broad-host-range plasmids such as RP4 permits the in vivo transfer of plasmids between nearly all species of Gram-negative bacteria [74]. Furthermore, it could be found that a rather small portion of genetic information has to be present in *cis* on a plasmid molecule. All other functions for conjugative plasmid transfer can be provided in *trans* on proper helper plasmids or by using host strains having the transfer genes integrated into the chromosome [75].

For example, a fragment of about 750 bp containing the origin of transfer-replication (*oriT*) of plasmid RP4 mediates mobilization of plasmids at high rates and even a 112 bp subfragment is sufficient for low frequency transfer [76, 77].

By using a transformable intermediate host (usually *E. coli* K12 strains) containing the transfer genes of a conjugation system on a helper plasmid or integrated into the chromosome, recombinant plasmids carrying a broad-host-range replication system and the *cis*-essential sequences of a conjugative transfer system can be mobilized into any target host [78]. When using strains with integrated *tra* genes or helper plasmids having a narrow host-range such as pRK2013 (*tra* genes of

RP4 fused to a ColE1 replicon)[79], or the narrow host-range *IncI* plasmids pLG221 and pLG223[80], only the recombinant broad-host-range plasmid will be maintained within the target cell.

Interspecies conjugative transfer could also be found in Gram-positive bacteria, mainly in streptomycetes[81,82] and in streptococci[83]. However, the mechanism of transfer seems to be completely different and the host-range might be even more restricted.

5 Vectors

Cloning vectors have to fulfil the basic requirements for the introduction of specific genes into a host organism. Two principal functions are therefore required of a vector molecule. These are:

1) To offer a useful selection for cells which have received a recombinant molecule.
2) To mediate the stable persistence of a gene fragment within a host organism.

In the case of shuttle vectors these basic requirements must be provided for both systems. In addition, the required elements for gene expression (promoters etc.) are in most cases integral parts of the vector systems.

5.1 Selection Systems

Since most of the DNA transfer systems are rather inefficient, it is necessary to have a suitable system available to select for clones carrying recombinant molecules. For most prokaryotic microorganisms dominant selection systems are available. These are mainly based on a great number of growth inhibiting antibiotics and on the corresponding resistance genes integrated into the vector molecules. The use of resistance to heavy metal ions has also been reported on[91].

In the case of eukaryotic microorganisms only few combinations of an antimicrobial active substance and a corresponding resistance gene are available. In some cases a bacterial resistance gene fused to eukaryotic expression signals can be used. However, such systems do not seem to be as universal as most of the dominant selection systems of prokaryotes.

On the other hand, a variety of selection systems based on complementation of specific auxotrophic mutations have been established. A great disadvantage of such systems is the necessity of having to isolate a specific mutant which can be used as a host cell for every organism. However, some of these systems were shown to be easily transferable to other species. Such systems are based on mutants which can be obtained very quickly and easily by using efficient, direct selection methods. For example, nitrate reductase deficient mutants can be obtained by selection for chlorate resistance[84]. Uracil auxotrophs having a mutation in the gene for orotidine-5'-monophosphate decarboxylase can be readily obtained by selection for 5-fluoroorotic acid resistance[85]. A gene fragment coding for this enzyme could be cloned from *Neurospora crassa* by using a selection based on complementation of *E. coli pyrF* mutants[86].

However, for the application of rDNA technology with industrial strains, dominant

Table 1. Selection systems for microorganisms

Microorganisms	Selectable genes	Remarks	Ref.
Prokaryotes			
Cyanobacteria	Cm^R, Em^R, Nm^R, Km^R, Sm^R	Heterologous genes originating from plasmids of Gram-negative and Gram-positive bacteria	88)
Gram-negative bacteria (various species)	Ap^R, Cm^R, Km^R, Sm^R, Sp^R, Su^R, Tc^R, As^R, Hg^R	Homologous and heterologous plasmid-bourne resistance genes	89–92)
Gram-positive bacteria			
Bacillus species	Cm^R, Em^R, Km^R, Sm^R, Tc^R, Tp^R	Resistance genes from *Bacillus* and *Staphylococcus aureus* plasmids	89–92)
	ilv, *leu*, *trp*	Complementation of auxotrophs by heterologous *Bacillus* genes	
Streptomycetes	Em^R, Nm^R, Hyg^R, $Thio^R$, Vio^R	Chromosomal genes from various *Streptomyces* species	61, 81, 95)
	Mel^P	Tyrosinase gene from *S. antibioticus*	
Corynebacteria	Cm^R, Em^R, Km^R, Nm^R, Tc^R, $Thio^R$, Sp^R	Resistance genes from *E. coli*, streptomycetes and *Streptococcus* species	96–99)
Streptococci, lactobacilli and clostridia	Em^R, Cm^R, Tc^R	Resistance genes from *Streptococcus* plasmids and from clostridia	83, 100–102)
Eukaryotes			
Yeasts			
Saccharomyces cerevisiae (laboratory and commercial strains)	*ARG4*, *HIS3*, *LEU2*, *URA3*, *TRP1*	Complementation of auxotrophs by cloned homologous genes	103–108)
	$G418^R$, Hyg^R, Cm^R	Bacterial resistance genes, prefarably fused to a yeast promoter	
	$Meth^R$	Mouse dihydrofolate reductase gene fused to a yeast promoter	
	CUP1	Gene from a copper resistant yeast strain	
Other yeasts	*HIS4*, *LEU*, *LEU2*	Complementation of auxotrophs by cloned homologous or by *S. cerevisiae* genes	109–113)
	$G418^R$	Bacterial resistance gene	
Myxomycota	$G418^R$	Bacterial resistance gene fused to a fungal promoter	114)
Filamentous fungi			
Zygomycota	*leu*	Complementation of auxotrophs by cloned homologous genes	115, 116)
	$G418^R$	Bacterial resistance gene	

Table 1. Continued

Microorganisms	Selectable genes	Remarks	Ref.
Ascomycetes and related fungi imperfecti	*qa-2*, *argB*, *met*, *ura-5*, *pyr-4*, *trp-1*, *trpC*	Complementation of auxotrophs by homologous or heterologous fungal genes	[117–129]
	amdS	Homologous or heterologous gene mediating growth on novel substrates	
	Hyg^R, $G418^R$, $Phleo^R$	Bacterial resistance genes fused to fungal promoters	
	Oli^R, Ben^R	Fungal genes cloned from resistant mutants	
Basidiomycetes	trp-1	Complementation of auxotrophs by cloned homologous genes	[130, 196]
Green algae	*ARG4*	Complementation of auxotrophs by heterologous (*S. cerevisiae*) genes	[131]

Abbreviations and genetic loci: Ap, ampicillin; As, arsenite; Ben, Benomyl; Cm, chloramphenicol; Em, erythromycin; G418, Geneticin; Hg, mercury; Hyg, hygromycin; Km, kanamycin; Mel, melanin; Meth, methotrexate; Nm, neomycin; Oli, oligomycin; Sm, streptomycin; Sp, spectinomycin; Su, sulfonamide; Tc, tetracycline; Thio, thiostrepton; Tp, trimethoprim; Vio, viomycin; *amdS*, acetamidase, *ARG4*, argininosuccinate lyase; *argB*, ornithine carbamoyl transferase; *CUP1*, gene locus mediating copper resistance; *HIS3*, imidazole glycerol phosphate dehydrogenase; *ilv*, isoleucine-valine biosynthesis; *leu*,*LEU*, leucine biosynthesis; *LEU2*, β-isopropylmalate dehydrogenase; *met*, methionine biosynthesis; *pyrG*, *pyr-4*, *URA3*, orotidine-5′-phosphate decarboxylase; *qa-2*, dehydroquinase; *trp*, tryptophan biosynthesis; *trp-1*,*TRP1*,*trpC*, trifunctional enzyme of tryptophan biosynthesis; *ura-5*, orotidylic acid phosphorylase; *P*, production; *R*, resistance

selection systems are necessary. Breeding of auxotrophic strains by mutation will generally result in a lowering of the productivity. With some systems, for example brewing yeasts, either asexual polyploid or aneuploid strains are used thus making the isolation of specific auxotrophic mutants almost impossible.

The use of genes mediating the utilization of specific substrates represents a further strategy for dominant selection. For eukaryotes, a useful system based on the utilization of acetamide or acrylamide mediated by the *amdS* gene of *Aspergillus nidulans* has been developed [87].

A collection of typical selection systems for various groups of microorganisms is shown in Table 1.

5.2 Types of Vectors

Two different mechanisms are available for the stable maintenance of recombinant DNA molecules within a host cell: 1) autonomous replication and 2) integration into the host genome.

For bacterial systems, autonomously replicating vectors have been mainly developed whereas for eukaryotic microorganisms integration seems to play a more important role.

5.2.1 Vectors for Bacteria

E. coli

A great diversity of useful vectors for general and for specific cloning purposes have been developed for the *E. coli* K12 system. Almost all of these vectors are based on autonomous replication provided by narrow-host-range replication systems of several plasmids and bacteriophages. These vectors can usually only replicate in *E. coli* or in very closely related enterobacteria.

A substantial part of the *E. coli* plasmid vectors is based on pBR322 derived multicopy plasmids carrying a ColEl related replication system. A review of pBR322 and its derivatives was recently given by Balbas et al. [132].

Several vector systems based on the bacteriophage λ have been mainly developed for use in the construction of genomic or cDNA gene libraries [133, 134]. Vectors based on the single-stranded bacteriophage M13 have been used to establish efficient DNA sequencing and site-specific mutagenesis strategies [135].

A very special group of vectors is based on a combination of plasmids and phages. These cosmids [136] contain the phage λ signal for packaging DNA into phage heads. Cosmids can carry large amounts of foreign DNA (up to 50 kb) thus they are mainly used to construct genomic libraries of higher eukaryotes [137].

A detailed compilation of *E. coli* and other cloning vectors was given by Pouwels et al. [138].

Other Gram-Negative Bacteria

Several naturally occurring plasmids characterized by a very broad host-range could be found in Gram-negative bacteria. These plasmids which belong to the incompatibility groups *IncP*, *IncQ*, and *IncW* have been used as a basis for the construction of various versatile cloning vectors able to replicate in many different Gram-negative genera including *E. coli* systems.

IncP cloning vectors are based on the conjugative plasmid RP4 or its identical but differently named isolates RK2, RP1 and R68. Several smaller derivatives have been constructed in vitro from this 60 kb plasmid [79, 92]. However, the very complex replication and maintenance system of RP4 creates problems since the deletion of some parts of RP4 results in lethal effects to the host cells [139]. An efficient system for the in vivo formation of deletion derivatives during conjugative transfer of RP4 was found which allowed the isolation of rather small derivatives [140].

The probably identical plasmids R300B and RSF1010 are the basis of *IncQ* vectors. With a size of 8.7 kb, these plasmids are much smaller than RP4. They do not possess a complete conjugation system but can be mobilized by helper systems carrying *tra* genes of *IncP* or $IncI_\alpha$ conjugative plasmids. The *IncQ* vectors are the best developed broad-host-range system and various derivatives are available for general and specific cloning purposes [89, 141–143].

Vectors based on the *IncW* plasmid pSa have been mainly constructed for use in *Agrobacterium* species [144, 145], but they are as well able to replicate in a variety of other Gram-negative bacteria.

Gram-Positive Bacteria

Well developed vector systems are available for *Bacillus* and *Streptomyces* species. In addition to plasmid vectors, bacteriophage vectors have been established with *Bacillus subtilis* [60] and with several *Streptomyces* species [146].

Plasmid vectors in *B. subtilis* are in most cases based on plasmids derived from *Staphylococcus aureus*. Several plasmids originating from thermophilic bacilli have also been used to develop *Bacillus* vector systems. It has been generally found that many of the plasmid vectors developed for *B. subtilis* can also be used for other *Bacillus* species. Descriptions of commonly used vectors for various *Bacillus* species are given in Refs. [60,146–150]. The construction of a vector system based on integration into the chromosomal DNA has also been reported for *B. subtilis* [151].

A large number of naturally occurring plasmids are known for *Streptomyces* strains and many of these are self-transmissible. Several of these plasmids have been used to construct various useful vectors [61,81,152] which usually can be used for different species of *Streptomyces*. Some combinations with *E. coli* plasmids have been constructed for shuttling genes between the two systems [153,154]. However, such vectors are not usually used for strain construction in *Streptomyces* species since plasmids containing *E. coli* sequences tend to be unstable in this genus [82].

Several host-vector systems have been developed for further Gram-positive bacteria of industrial interest. For instance, in the case of the amino acid producers *Brevibacterium lactofermentum* and *Corynebacterium glutamicum*, well developed systems are available [155]. Examples of vectors for corynebacteria and some other Gram-positive bacteria of industrial interest are listed in Table 2.

Table 2. Selection of cloning vectors for Gram-positive bacteria of industrial interest

Organism	Vector	Source of replicons	Selectable markers	Ref.
Brevibacterium	pUL62	pBL1[c] (pBR322)[a]	Km^R (Ap^R, Km^R)[a]	156)
lactofermentum	pUL340	pBL1[c]	Km^R (Km^R)[a]	157)
	pAJ224	pAM330[c]	Tp^R	158)
	pGX1415	pGX1901[c] (pUB110)[b]	Km^R (Km^R)[b]	159)
	pGX1418	pGX1901[c] (pUB110)[b]	Km^R, Cm^R (Km^R, Cm^R)[b]	159)
Corynebacterium	pHY416	pSR1[d] (pBD10)[b]	Km^R (Km^R, Cm^R)[b]	96)
glutamicum	pCS11	pCG1[d]	Sp^R, Sm^R	160)
	pCE53	pCG1[d] (pGA22)[a]	Km^R, Cm^R, Tc^R (Km^R, Cm^R, Ap^R, Tc^R)[a]	98)
Brevibacterium	pAJ655,	pAM330[c] (pBR325)[a]	Cm^R (Cm^R, Ap^R)[a]	161)
lactofermentum and	pAJ43			
Corynebacterium	pAJ440	pAM330[c] (pUB110)[b]	Km^R (Km^R)[b]	161)
glutamicum	pSA77	pX18[c] (pBR325)[a]	Em^R, Cm^R (Ap^R, Cm^R)[a]	99)
Streptococcus lactis	pGK12	pWV01[e]	Cm^R, Em^R (Cm^R, Em^R)[a, b]	162)
Streptococcus sanguis	pVA838	pVA380-1[f] (pACYC184)[a]	Em^R (Em^R, Cm^R)[a]	163)
Lactobacillus plantarum and several other lactobacilli	pAMβ1*	pAMβ1[g]	Em^R	100)
Clostridium	pAMβ1*	pAMβ1[g]	Em^R	83)
acetobutylicum	pVA797*	pIP501[h]/pVA380-1[f]	Cm^R (Cm^R)[a]	164)
	pAM610*	pMV163[h] (pMK20)[a]	Km^R, Tc^R (Km^R, Tc^R)[a]	164, 165)

The abbreviations used for markers are given in the footnote of Table 1. Replicons and Selection markers in brackets refer to shuttle organisms. The origin of the replicons and the shuttle hosts refer to the following organisms: a *Escherichia coli*, b *Bacillus subtilis*, c *Brevibacterium lactofermentum*, d *Corynebacterium glutamicum*, e *Streptococcus cremoris*, f *Streptococcus ferus*, g *Streptococcus faecalis*, h *Streptococcus algalactiae*. Conjugative or mobilizable plasmids are marked by an asterisk *

5.2.2 Vectors for Eukaryotic Microorganisms

With respect to the maintenance of vectors, a completely different situation is present in eukaryotes. Autonomously replicating plasmid systems are rare and with some exceptions are predominantly confined to *Saccharomyces cerevisiae* and related yeasts.

Yeasts

The replication and maintenance system of the natural 2μ plasmid of *S. cerevisiae* and cloned chromosomal "autonomous replicating sequences" (*ARS*) have been used to construct a variety of replicating plasmid vectors for yeasts. Any circular plasmids without a replication system, but containing homologous sequences to chromosomal genes, can in principle be used as integrative vectors. However, transformation frequencies are usually much lower in comparison to replicating vectors. In addition to plasmid vectors, artificial linear chromosomes containing *ARS* sequences, centromeres and telomeres have also been developed for use as vectors [166, 167]. However, at the present time these vectors are only of scientific interest. In addition it has also been shown that small minichromosomes are characterized by a reduced mitotic and meiotic stability [168, 169]. Descriptions of commonly used vector systems for *S. cerevisiae* are to be found in Refs. [62, 103, 170].

Vector systems for other yeasts are mainly based on cloned homologous *ARS* sequences. However, in several yeasts vectors based on heterologous *ARS* sequences (*S. cerevisiae*) or on the 2μ replication system also function. For example, replicative vector systems have been described for *Schizosaccharomyces pombe* [111], *Kluyveromyces lactis* [171, 172], *Kluyveromyces fragilis* [112], *Candida maltosa* [110], *Pichia pastoris* [109], and some further strains. An integrative vector system based on recombinational integration via homologous sequences has been developed for *Candida albicans* [173].

The vectors developed for laboratory strains of *S. cerevisiae* can in principle be used for industrial strains of *S. cerevisiae* (brewers and bakers yeasts, wine yeasts). However, the relevant problems are more closely related to finding a proper selection system. Furthermore, integrative vectors are desired for strain improvement purposes because of the instability of autonomous replicating vectors [174].

Filamentous Fungi

Within the last few years much effort was expended for the development of suitable host-vector systems for various fungal species which are primarily of substantial industrial importance.

In filamentous fungi only a few circular plasmids of mitochondrial origin and several linear plasmids have been found and little is known at present about these elements [175]. The use of fungal plasmids as a basis for vector development has been reported only for *Neurospora crassa* [176] and for *Podospora anserina* [177]. In addition, "autonomous replicating sequences" cloned from fungal chromosomal or mitochondrial DNA in most cases turned out to be functional in the yeast *S. cerevisiae* (used for selection of *ARS* sequences), but not in the original organisms [178–180]. However, some exceptions are known, mainly in the case of Zygomycotina where homologous *ARS* sequences seem to function [115, 116, 181].

Table 3. Compilation of cloning vectors for filamentous fungi

Fungus	Vector[5]	Selectable markers[1]	Transformation[2] frequencies	Remarks	Refs.
Zygomycetes					
Mucor circinelloides	pMCL1302[r]	*leu* (*TRP1*, *URA3*)[b] (Ap^R)[a]	7.800 (3×10^3)[3]	Contains ARS sequence from *S. cerevisiae*	115, 181)
	pMCL002[r]	*leu* (Ap^R)[a]	7.800 (3×10^3)[3]	No yeast sequences present	
Phycomyces blakesleeanus	pJL2[r]	G418[R h] (URA3)[b] (Ap^R, Km^R)[a]	40–60	Contains a *Phycomyces* derived *ARS* sequence	116)
Ascomycetes					
Aspergillus awamori	n.d.	*pyr-4*	n.d.	Referred to unpubl. results	184)
Aspergillus nidulans	p3SR2[i]	*amdS* (Ap^R)[a]	300–400	*amdS* deletion strains used as host organisms	185, 186)
	pSa143[i]	*argB* (LEU2)[b] (Ap^R)[a]	50		121)
	pHY201[i]	*trpC* (Ap^R, Cm^R)[a]	20–150		119, 180)
	pDJB1[i]	*pyr-4*[c] (Ap^R, Cm^R)[a]	85–200		180, 187)
	pDJB2[i]	*pyr-4*[c] (Ap^R)[a]	3000–6000	*ANS1* sequence inserted into pDJB1	180, 187)
	pKBY2[i]	*trpC* (Ap^R, Cm^R)[a]	10^4	Can be used for cosmid cloning in *E. coli*	188)
	pAN7-1[i]	Hyg[R g] (Ap^R)[a]	5–15		363)
Aspergillus niger	pDG3[i]	*argB*[d] (Ap^R, Km^R)[a]	4		122)
	p3SR2[i]	*amdS*[d] (Ap^R)[a]	4	No detectable homology of *amdS* in *A. niger* DNA	87)
	pAB4-1[i]	*pyrG* (Ap^R)[a]	40		85)
	pAN7-1[i]	Hyg[R g] (Ap^R)[a]	10–20		363)
Aspergillus oryzae	pKA5-1[r]	*met* (*LEU2*)[b] (Ap^R)[a]	14–32	Contains sequences of the yeast 2μ plasmid	124)
Cephalosporium acremonium	pIT221[i]	Hyg[R c] (Ap^R)[a]	5		125)
	pPS21[i]	Hyg[R g] (Ap^R)[a]	250		364)
	pMA3, pMA4[i]	G418[R f] (Ap^R, Nm^R)[a]	0.2–1	Only 20–50 % of resistant colonies represent transformants.	126)
	Unnamed[i]	G418[R g]	0.3	Most resistant colonies represent transformants	126)

Table 3. Continued

Cochliobolus heterostrophus	p3SR2[i]	*amdS*[d], $(Ap^R)^a$	0.02–0.06	No detectable homology of *amdS* in *C. heterostrophus* DNA	189)
Neurospora crassa	pVK88[i]	*qa-2*	5–30		117)
	pALS1[r]	*qa-2*, $(Cm^R)^a$	100–260 }	Contain sequences of a mitochondrial plasmid, rearrangements in *N. crassa*	175)
	pMK2[i]	*qa-2*, $(Tc^R)^a$	140 }		190)
	pRAL1[r]	*qa-2*, $(Tc^R, Cm^R)^a$	10,000–50,000	High frequencies obtained with a specific lot of Novozyme 234	191)
	pNC2[i]	*trp-1*, $(Ap^R)^a$	170–500		118)
	pJR2[r]	*am*[h], $(Ap^R)^a$	up to 12,000	Multimere formation of vector in fungal host	192)
Penicillium chrysogenum	pPCtrpC1[i]	*trpC*, $(Ap^R)^a$	40–80		193)
	p3SR2[i]	*amdS*, $(Ap^R)^a$	2–20; up to 120		127, 194)
	pJDB2[i]	*pyr-4*[c]	1,200–3,000	Contains *ANS1* from *A. nidulans*	123)
	pHS103[i]	Phleo[R g], $(Ap^R)^a$	5–15		127)
Podospora anserina	pPAura5[i]	*ura-5*, $(Ap^R)^a$	1–2		120)
	Unnamed[i]	*leu*	10–20	Transforming plasmid contains a suppressor tRNA gene	195)
Trichoderma reesei	n.d.	*pyr-4*[c]	n.d.	Referred to a pers. commun.	184)
	p3SR2[i]	*amdS*[d] $(Ap^R)^a$	50–100		365)
	pSa143[i]	*argB*[d] $(Ap^R)^a$	140–400		365)
Basidiomycetes					
Schizophyllum commune	pAM1[i]	*TRP1* $(Ap^R)^a$	3–30		130)
Coprinus cinereus	pCc1001[i] pCc1002[i]	*TRP1*, $(Ap^R)^a$	$(1 \cdot 10^{-4})^3$		196)
Ustilago maydis	pMP81	Nm[R f]	1–10	Unstable transformants, contains yeast 2µ DNA	197)

The abbreviations used for markers are given in the footnote to Table 1. Selection markers in brackets refer to shuttle organisms. 1 no indication means the use of homologous gene fragments. Heterologous gene fragments are originating as follows: a *Escherichia coli*, b *Saccharomyces cerevisiae*, c *Neurospora crassa*. d *Aspergillus nidulans*, e bacterial gene fused to a yeast promoter, f bacterial gene with native promoter, g bacterial gene fused to a fungal promoter. h glutamate dehydrogenase gene of *Neurospora crassa*. 2 Transformation frequencies are usually given as the number of transformants per µg vector DNA. 3 Transformation frequency given as number of transformants per viable protoplasts. 4 Transformation frequency refers to cosmid clones containing inserts of 30–40 kb. 5 The maintenance system of the vector in the fungal hosts as specified by the authors is as follows: i integration into the host genome, r autonomous replication. n.d.: no data available

A sequence enhancing the transformation efficiency of an integrative vector could be isolated from *Aspergillus nidulans* chromosomal DNA [180]. Although this sequence was isolated on the basis of replicating activity in *S. cerevisiae*, there is no evidence for autonomous replication in *A. nidulans*. The function of this sequence (*ANS1*) still remains unclear.

The lack of autonomous replication systems in most filamentous fungi is not a problem from the viewpoint of strain improvement since the known eukaryotic replicating plasmid systems are more or less rather unstable. It has already been shown with yeast vectors that integration into the host chromosome results in a high mitotic and meiotic stability. Therefore, despite low transformation frequencies, integrative vectors represent an adequate alternative for fungal systems. Furthermore, integration of DNA at non-homologous sites seem to be a frequent event in filamentous fungi even allowing the use of vectors without extended homology to any region of the fungal genome [183]. In addition, in many cases more than one copy of a vector is integrated thus representing a possibility for gene amplification.

Various vectors have now been developed for filamentous fungi and several examples of these are listed in Table 3.

Green Algae

Autonomously replicating vectors could be constructed for the unicellular green alga *Chlamydomonas reinhardii*. Four different small DNA fragments mediating *ARS* activity could be isolated in a shotgun cloning experiment using the total DNA of this organism. Interestingly enough, all four elements turned out to be part of the chloroplast DNA. One of these fragments shows *ARS* activity in yeast [198].

6 Expression of Cloned Genes

6.1 Basic Aspects

Much work has been done in the last years on the analysis of gene expression and its regulation in prokaryotes and eukaryotes. For instance, in the case of *E. coli* the basic genetic elements involved in gene expression are now much better understood. These basic genetic elements are the following:

1) Promoter regions consisting of the binding sites for RNA polymerase (—10 region) and for the sigma factor (—35 region), of binding sites for regulatory proteins (operator, CAP site) and the transcription start site [199–201].
2) Transcription terminators [202].
3) Translation signals including ribosome binding sites and start and stop codons [203].

However, even with the *E. coli* system, the concepts pertaining to the structure of expression signals are still developing. For example, it is now well documented that further classes of promoter elements can be present in distinct groups of genes and that sequences upstream and downstream from the classical promoter elements (—10 and —35 region) may have a significant influence on promoter activity [204–206]. Moreover, there is some evidence that the general sequence structure of a gene will have an influence on expression activity. It has been well established for a number of different organisms that promoters may be gene-specific with respect to the production of appropriate mRNA levels [207]. Such effects are not well understood at the present time and these are some of the reasons why the development of several genetically engineered pharmaceuticals has not been that rapidly successful as might have been hoped for [208].

In addition to these primary requirements, a variety of secondary factors affecting gene expression have to be taken into account. This is of particular significance when heterologous genes must be expressed. Since many of these factors are poorly understood, they may cause severe problems. The most important of such factors are listed in Table 4.

Table 4. Secondary factors affecting gene expression

1)	Correct and efficient RNA processing
2)	mRNA stability and secondary structure (loop back)
3)	Codon usage
4)	Protein processing and post-translational modifications
5)	Formation of correct three-dimensional protein structure (folding, disulfide bonds)
6)	Correct assembling of protein subunits
7)	Toxicity of gene products to host cells
8)	Instability of gene products in the host system used (e.g. degradation of proteins)

In addition to the *E. coli* system, prokaryotic gene expression signals have been better studied in *Bacillus* [209] and in *Streptomyces* species [210, 211]. In general, the expression systems of Gram-positive and Gram-negative bacteria are closely related. However, there exist some differences in the structure of promoters which might provide barriers to heterologous expression of genes. For example, very few nucleotide sequences that serve as promoters in *Streptomyces* can function in *E. coli* [212], but many *E. coli* promoters have activity in *Streptomyces* [213]. In contrast to this, *Bacillus* promoters are usually recognized in *E. coli* whereas most *E. coli* promoters have no activity in *B. subtilis* [214].

In comparison to prokaryotes, much less information on gene expression and its regulation is available for eukaryotic systems. In general, the situation is much more complex and it is now clear that in addition to sequence determined elements (promoters, terminators, ribosome binding sites etc.) other factors such as chromatin structure and molecular environment may have a great impact on expression activity [215].

Among the eukaryotic microorganisms, yeast gene expression systems have been best studied. Unlike the situation in *E. coli*, transcription is carried out by three different RNA polymerases. RNA polymerases I and III transcribe rRNA and tRNA genes whereas RNA polymerase II is the enzyme which transcribes all the protein coding genes. A further difference is that transcription units of more than one gene (operons) do not seem to exist in eukaryotic systems.

Yeast RNA polymerase II promoter regions are large when compared to prokaryotic systems and furthermore, for various genes they differ considerably in size. The size range is between 80 and 500 base pairs [216, 217]. The general features of yeast promoter elements are shown in Fig. 2.

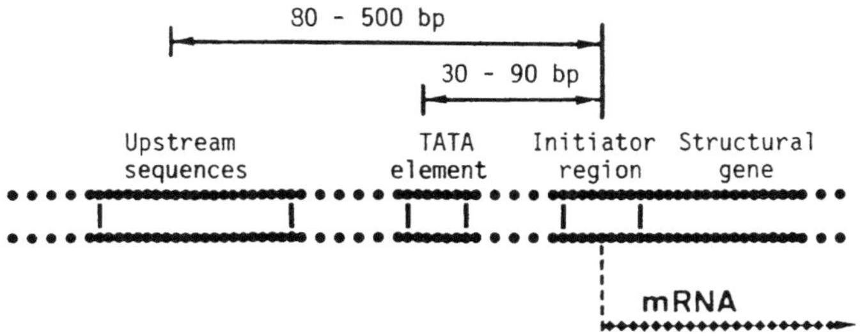

Fig. 2. Promoter elements in yeast. The scheme is based on data presented in a recent review on yeast promoters [217]. The initiation region directs the start site for transcription. The "TATA" element is characterized by a short (dA-dT) region which seems to be involved in transcription initiation. It is usually located between 30 to 90 base pairs upstream from the transcription start site. Multiple, independent "TATA" sequences can be present which may specify different transcription start sites [218,219]. However, in higher eukaryotes a similar "TATA" element is invariably located 25–30 base pairs from the transcription initiation site. The upstream elements are usually specified by short sequences involved in activation or repression of transcription activity [216,220]. Such sequences can be located up to 500 base pairs upstream from the transcription start. Multiple, distinctly regulated upstream sites have been described to be present in several genes [221,222]. Poly (dA-dT) sequences were found to be upstream promoter elements for constitutive expression in yeast [223]

At the present time, only a few genes of filamentous fungi have been characterized with respect to their promoter regions. The existing results indicate that the basic structures are similar to those of yeast promoter regions [224,225]. However, the "TATA" region seems to be less conserved, at least in the case of genes of *Neurospora crassa* [226].

6.2 Control of Gene Expression

Soon after the first cloned eukaryotic genes were expressed in *E. coli*, it became clear that the connection of a foreign gene to a strong expression system in itself will not sufficiently solve the expression problem. Uncontrolled overexpression of a (foreign) gene can inhibit cell growth or even kill the host cells. In several cases good results could be obtained by the use of tightly regulated promoters which can be turned on only when required cell densities are reached [208].

In view of strain improvement by rDNA techniques, controlled expression of specific genes will be a dominant aspect. Therefore, the availability of well characterized regulation systems is of great importance.

For prokaryotes there exist well founded concepts about gene regulation via negative and positive control mechanisms [227]. In addition to these basic mechanisms, some different regulatory systems have been identified. The translation coupled attenuation mechanism [228] or the antisense RNA regulation [229] have been already well documented. In addition, regulation of protein synthesis at the level of translation has been also reported [230].

The majority of the presently used *E. coli* expression vectors is based on three promoters: *lambda* P_L, *trp*, and *lacUV5* [231]. All three promoters are considered as being strong, can be repressed and turned on, and may direct the expression of a foreign gene to levels of up to 40% of the total cellular protein. However, it is not easy to achieve regulation of these promoters to a distinct level of activity at any stage of cell growth.

Gene amplification has been generally found to provide a possibility of enhancing gene expression [232]. Vectors based on temperature-dependent runaway-replication mutants of plasmids offer the possibility of controlled elevation of the copy number of a cloned gene [233]. The combination of regulated promoters together with the priming system for plasmid replication represents an improved system which allows the adjustment of the copy number of a plasmid at any distinct level [231, 234]. Vectors containing such a regulated replication system may, therefore, be a good basis for the development of expression systems which can be tuned very precisely.

One main feature of the regulation of gene expression in eukaryotic microorganisms is the predominance of positive control mechanisms. The *cis* acting regulatory target sites are connected to the upstream promoter elements. However, fine structure mapping indicates that for some genes the upstream promoter elements and the regulatory sites are distinct and separable DNA sequences [217].

In the case of yeast, inducible expression systems are poorly developed at the present time. Most reports describe the use of the *GAL1* upstream region which mediates up to 2000-fold induction by galactose, and the *PHO5* promoter which is induced about 200-fold by phosphate starvation [44]. Recently, the *ADHII* promoter which is repressed at high glucose concentrations and switched on at a low glucose level was discussed as being an inherently promising system [235].

Several genes involved in either positive or negative regulation have already been described for filamentous fungi [236, 237]. A system which has been better characterized

Table 5. Expression of an *Escherichia coli lacZ* fusion gene in *Penicillium chrysogenum*

Strain	Colour on X-gal plates	β-Gal activity (U mg^{-1})[a]	Copy number[b]
Tr.-1	Blue	8783	2
Tr.-2	Blue	16238	5
Tr.-3	Blue	11613	> 6
Tr.-4	Light blue	3860	>10
Tr.-6	White	0	0
Tr.-8	Light blue	8293	> 6
P2	White	0	0

Plasmid pAN5-41B carrying an *E. coli lacZ* gene fused in phase to the 5′-region of the *Aspergillus nidulans* glyceraldehyd-3-phosphate dehydrogenase gene [239] was introduced into *P. chrysogenum* strain P2 by cotransformation with an *amdS* vector [127]. Transformants selected for growth on acetamide or acrylamide were tested for expression of the *lacZ* gene on media containing X-gal. a: Cell lysates were prepared from mycelia powdered under liquid nitrogen and β-galactosidase activity was determined using a standard colorimetric assay. b: DNA of transformants was digested with *BamHI* and subjected to a Southern analysis using a radiolabeled *lacZ* gene fragment. The number of bands on the Southern blot is an indication for the copy number of integrated *lacZ* sequences is the *alcR* gene of *Aspergillus nidulans*. It exerts a positive control on the genes *alcA* and *aldA*, and also a negative control on the gene *alcB*. *alcR* is closely linked to *alcA*, but unlinked to *aldA* and *alcB* [238]. The *alcA* promoter has been used to express the human interferon alpha 2 gene in *A. nidulans*. It turned out that when multiple copies of the fusion gene were integrated into the host genome, expression levels increased only to a certain plateau which was caused by a lack of sufficient *alcR* gene product. The additional introduction of multiple copies of the *alcR* gene resulted in higher expression levels [235].

By using the *E. coli lacZ* gene fused to 5'-regions of fungal genes, quantitative studies of the expression levels of such fusions in a fungal host could be performed [239]. For example, a fusion gene of *E. coli lacZ* and the *A. nidulans gpd* 5'-region was transformed into *Penicillium chrysogenum*. The analysis of the expression levels in several transformants showed that in some cases there is no correlation between expression levels and copy number of integrated fusion genes (Table 5). This indicates that gene expression levels may be influenced by the position of the integrated *lacZ* fusion gene [127].

6.3 Protein Secretion

One domain for the application of rDNA techniques in biotechnology is the production of polypeptides. During the past few years much emphasis has been put on the production of mammalian proteins by microorganisms for pharmaceutical purposes. Considerable effort has been expended on the development of systems for *E. coli*, yeast and for *B. subtilis* which would allow excretion of the desired polypeptides into the medium.

It is in general a well known fact that N-terminal signal peptides are shared by all exported proteins [240]. Fusion of a foreign gene to the C-terminus of a signal peptide has been shown in several cases to permit secretion and proper cleavage at the fusion site [43, 241–243]. However, the efficiency was found to be low in most cases even with organisms that naturally excrete large amounts of proteins such as *B. subtilis*. Among the many reasons for this that have been discussed, two seem to have a more general implication:

1) Only limited amounts of protein can be secreted due to limited expression of enzymes involved in the secretion process (e.g. signal peptidase) [243].
2) There is considerable evidence that efficient secretion can only be obtained with proteins that are secretory in their original environment [244].

It is thus evident that at the present time there is a considerable lack of knowledge of the genetics and molecular biology of secretion systems. The increasing potential of basic research on secretion will possibly provide an improved basis for the development of more efficient secretion systems.

Much attention has been devoted to filamentous fungi since certain species can secrete large amounts of protein and the first results seem to be quite promising. The signal sequence of a glucoamylase gene from *Aspergillus awamori* can direct the secretion of glucoamylase from yeast [245]. Recently, the secretion of bovine chymosin from *A. nidulans* using the transcriptional, translational, and secretory control units of the *Aspergillus niger* glucoamylase gene has been reported [246].

Filamentous fungi are not only of interest for the production of proteins of higher eukaryotes. Various species of this group of microorganisms are already in use in large scale industrial processes for the production of commercial enzymes [247]. In nearly all of these processes extracellular enzymes are produced. Therefore, the development of efficient secretion systems with filamentous fungi will be of great importance for future strain improvement strategies in the biotechnology industry.

7 Industrial Processes with Recombinant Microorganisms

Despite the voluminous scientific literature which has evolved since 1975 in genetic engineering, only a few publications pertain to the impacts of molecular biology on process research and development. Of the nearly 250 papers published in a leading journal in biochemical engineering in 1985, only six papers described research involving recombinant microorganisms [248]. A possible explanation for this is the fact that currently most practical applications of rDNA are connected with the production of polypeptides of pharmaceutical interest. These "high price — low volume" products are usually produced under conditions which are comparable to laboratory-scale experiments. Relatively small volumes of defined media are usually used. A second factor is that many of the developments involving recombinant microorganisms only exist presently at the laboratory level and large-scale production processes are only instigated later following further development of the process.

In general, problems associated with scaling up of bioprocesses employing recombinant microorganisms should be similar to those normally encountered in processes using strains developed by classical methods. However, depending on the host-vector system used, some specific problems discussed in the following may have a substantial impact on process development strategies.

7.1 Genetic Instability

The genomes of microorganisms are for the most part extremely stable. However, rare hot spots for instability exist in particular organisms, genomes or genes. The effects of instability include for example, loss or destruction of whole linkage groups (chromosomes, plasmids, organelle genomes) and major DNA rearrangements and transpositions [249]. Another well-known fact is that highly productive strains obtained from genetic breeding programs degenerate very rapidly upon subculturing, thus resulting in a serious decrease of product yields [250]. The main reason for this instability can be attributed to the fact that highly productive strains have been genetically tailored to overproduce a particular substance and thusly growth and/or other reproductive sequences such as sporulation are negatively influenced [251]. Cells which undergo genetic changes resulting in lower productivity are usually characterized by an increased vigour and their progeny will dominate the population after a few generations.

Stable maintenance is also one of the central problems of large-scale bioprocesses employing recombinant microorganisms. Two categories of instability can be defined: 1) segregational instability and 2) structural instability.

7.1.1 Segregational Instability of Autonomously Replicating Plasmid Vectors

At present, vectors based on autonomously replicating plasmids are the systems predominantly used for bacteria and yeasts. The stable propagation of a plasmid requires efficient replication and accurate partitioning of plasmid molecules to daughter cells during cell division [252]. Naturally occurring plasmids are very stable. However, the deletion derivatives commonly used as vectors for rDNA research and applications are often much less stable [253, 254]. In addition, cells containing recombinant plasmids usually show a reduced fitness in comparison to their plasmid-free homologs [253]. Moreover, the vigour of plasmid-carrying cells can be drastically reduced by overexpression of plasmid-encoded genes [255–257]. These problems can be overcome in principle by applying a selective pressure against plasmid-free cells. The most commonly used selection systems are based on antibiotic resistance. However, this is a practicable strategy in the laboratory but for several reasons it is not applicable to large-scale processes.

One main reason for segregational instability of plasmid vectors is the absence of proper partitioning systems which were eliminated during the construction of vectors from natural plasmids. Random partitioning may be sufficient to ensure stable inheritance of high copy plasmids but an active partitioning system is necessary for low copy plasmids [258].

Gene regions coding for partitioning systems (*par*) have been identified in the case of plasmids R1 [259], F [260], P1 [261], NR1 [262], pSC101 [263], and RP4 [264]. In

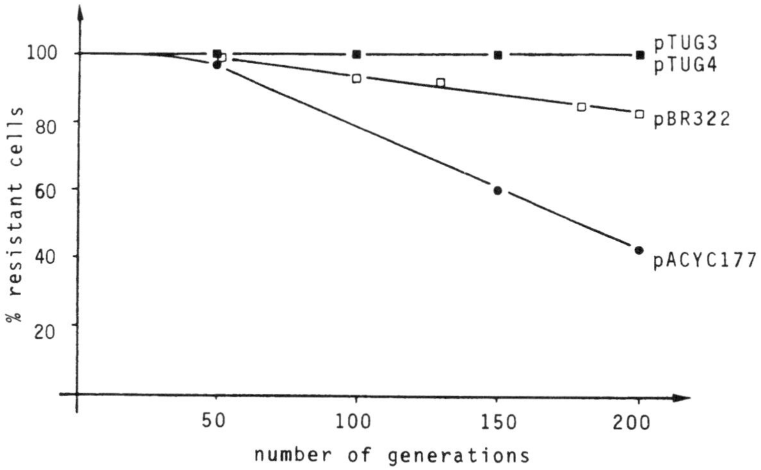

Fig. 3. Stabilization of plasmid vectors by the *par*-region of plasmid RP4. The *par*-region of the broad-host-range plasmid RP4 [264] was cloned into the *Escherichia coli* vector plasmids pBR332 and pACYC177. The resulting plasmids are pTUG3 and pTUG4, respectively. Plasmid loss under non-selective conditions was determined in batch culture. *E. coli* strains carrying the respective plasmids were grown overnight in a selective medium (generation 0). The cultures were then diluted 1 : 1000 into a non-selective medium and grown for 8 h at 37 °C to the early stationary phase. This cycle was repeated until the desired number of generations was attained. One cycle corresponds to nearly exactly 10 generations. Samples were taken from several cycles and analyzed for the loss of resistance markers by replica plating on non-selective and selective media

addition, high copy plasmids may also carry gene functions which contribute to plasmid stability. Examples of this are found with plasmids ColE1 [258] and CloDF13 [265]. Interestingly, the eukaryotic 2μ plasmid of yeast also contains a stabilizing gene function which is comparable to a *par* system of bacterial plasmids [266].

The molecular details of the mechanisms mediated by such stabilizing gene regions are still not well understood. Site specific attachment of plasmid molecules to specific membrane sites which provide segregation of at least one copy into the newly formed daughter cells is the most popular model. For the R1 *par* system some evidence for the presence of such a mechanism could be provided [267]. However, for the stabilizing loci identified in ColE1 and CloDF13, site specific recombination systems which resolve plasmid multimers to monomers thus maximizing copy number and stability have been described [258, 265].

On the other hand, partitioning functions of low copy plasmids were shown to stabilize high copy plasmids [264, 268]. An example for the stabilizing effect of the *par* region of plasmid RP4 on standard *E. coli* vectors is shown in Fig. 3.

Nearly all of the studies on the stabilizing effects of *par* regions of bacterial plasmids have been carried out with the *E. coli* system. It could be shown only for the RP4 *par* region that it is functional in several Gram-negative bacteria including *Alcaligenes*, *Azotobacter*, *Klebsiella* and *Pseudomonas* species [264].

In addition to the true partitioning systems of low-copy plasmids which presumably act by distributing the plasmid molecules to the daughter cells, further loci involved in decreasing the occurrence of plasmid-free cells have been described. The *ccd* mechanism of the F plasmid and the *hok* gene of plasmid R1 seem to kill those cells postsegregationally which did not obtain a copy of the respective plasmids [269, 270].

Partitioning regions cloned into recombinant plasmids have already been shown to be suitable for enhancing the stability of recombinant plasmids constructed for production purposes [271–273]. In some cases the application of a selective pressure which eliminates plasmid-free cells will have an advantage compared to the use of a partitioning system [273]. Several systems which can be used on an industrial scale have been developed. They are mainly based on the cloning of a gene on a plasmid vector which is needed for growth by the host cell. A common system is the use of an auxotrophic mutant and a cloned gene complementing the deficiency [273, 274]. However, the introduction of auxotrophic markers usually reduces the overall efficiency of a cell and furthermore, minimal media must be used. This again is not a desirable feature for industrial scale processes. A double mutant system has been described for yeast which allows the use of a complete medium. In addition to the mutation leading to auxotrophy, a second mutation blocks the active transport of the required metabolite into the cell [275].

Selective pressure for plasmid containing cells could also be achieved by the use of genes coding for toxic bacteriocins [276] and the use of systems based on the induction of lysogenic phages in plasmid free cells [277]. Genes cloned into recombinant plasmids coding for the utilization of a specific substrate which is used as a main nutrient source would be an alternative strategy.

Stabilization of autonomously replicating eukaryotic vectors could be achieved by combining them with a centromere region of a chromosome. Since centromeres are responsible for the exact partitioning of eukaryotic chromosomes during mitosis and meiosis, they would be expected to mediate high stability. However, such expectations were not completely fulfilled. Although a certain stabilizing effect on yeast *ARS* plasmids could be found, the rate of improper segregation was much higher than observed for a normal chromosome [278].

7.1.2 Structural Instability in Recombinant Organisms

Although the problems caused by segregational instability are predominant with autonomously replicating vectors, there also are several reports on structural instability. In Gram-negative bacterial hosts, structural instability was found to be mainly connected with drastic overexpression or with toxicity of gene products. As a consequence of this, structural instability is generally characterized by deletions, insertions, rearrangements, point mutations etc. which eliminates (parts of) the responsible genes or reduce the level of expression [256, 279, 280]. There is some evidence for the existence of "hot spots" for instability and the involvement of *IS* elements has been found in some cases [281]. Transposons and insertion elements are general sources of genetic variability [282].

In Gram-positive bacteria structural instability is a more serious problem. Genomic instability seems to be a common phenomenon in *Streptomyces* [283]. Deletion, amplification and rearrangement of large DNA segments was reported to occur at high frequencies [284–286]. Structural rearrangements of recombinant plasmids in *Bacillus subtilis* have been described in a number of studies [287, 288]. It was shown that deletions occur most frequently by recombination between short homologous sequences and it is suggested that single stranded replication intermediates are responsible for this [289]. Presently there are no well-founded strategies to overcome such problems because of a very poor understanding of the factors causing structural instability.

For eukaryotic microorganisms there is a clear trend to the use of integrative vectors which avoid the segregative instability problem. Yeast integrative vectors based on homologous integration are generally considered to be very stable. However, detailed studies on stability seem to be rare. One report on the stability of a chloramphenicol resistance vector integrated via the *HIS3* gene in several commercial yeast strains describes 100% maintenance of the marker gene under non-selective conditions of the industrial process [107].

Several studies on the stability of integrated sequences have been carried out with filamentous fungi. In *argB* transformants of *Aspergillus nidulans*, mitotic stability was usually found to be very high while most transformants showed low levels of meiotic instability [121, 290]. A similar situation was found with *amdS* and *trpC* transformants of *A. nidulans* [119, 185, 186, 291]. Many further reports on transformation of filamentous fungi generally show that integrated sequences remain quite stable at least during mitosis. Since nonhomologous and multiple (tandem) integration is a frequent event in filamentous fungi, one would expect genetic instability in such transformants. Indeed, for *A. nidulans* transformants having integrated multiple copies of pBR322 containing *A. nidulans* sequences of the *prn* gene cluster, the generation of deletions by recombinational events was found [292]. However, the situation with integrated vectors seems in general to be quite satisfactory for all practical purposes.

7.2 Process Conditions and Problems of Scaling-Up

The stable maintenance of engineered DNA sequences in a production organism represents a central problem which is also related to many aspects of process conditions and scale-up. However, the importance of environmental conditions for optimizing (recombinant) gene expression has also been recognized.

The influence of nutrient limitation on plasmid stability has been investigated with several host-vector systems [254, 255, 293, 294]. Experiments were performed in carbon-, nitrogen-, sulfate-, phosphate-, magnesium-, and potassium-limited chemostats. The majority of the plasmids exhibited the highest instability under phosphate limitation. Plasmid stability was also found to be dependent on the growth rate. In batch cultures, plasmid loss increases in the stationary phase [295]. Chemostat studies also revealed that plasmid stability is reduced at lower dilution rates [296, 297]. The influence of temperature on the stability and on the growth kinetics of plasmid carrying cells was investigated in *E. coli* and *Pseudomonas* hosts. At a lower temperature (30 °C), growth kinetics were not affected by the presence of plasmids but plasmid instability was observed. At a higher temperature (37 °C), growth rates and yields were lower than for plasmid-free hosts, but the plasmid stability was high [298].

Based on such investigations and on theoretical considerations, various mathematical models for describing the growth and behaviour of recombinant microorganisms and strategies for optimal process design have been developed [299–309]. A condensation of these models and strategies is presented in a recent review [248]. However, with respect to production processes, experimental data which would verify the usefullness of the proposed models and strategies are rare. Some data on the production of proteins with genetically engineered *E. coli* are to be found in Refs. [248, 310, 311].

At the present time sufficient data is not available in order to permit a detailed discussion of the scaling-up problems with processes employing recombinant microorganisms. However, it should be obvious from the previously discussed facts on stability and gene expression that recombinant microorganisms require well defined environmental conditions to attain optimal productivity. Concentration or temperature gradients may have an enormous influence. Therefore, all aspects of heat and mass transfer play an important role in scale-up strategies [312]. There is at least one report which shows that a shock with respect to the dissolved oxygen concentration causes instability in *E. coli* containing recombinant plasmids [313].

8 Strategies for Strain Improvement by rDNA Technology

8.1 General Aspects

The main targets of strain improvement programs are to increase the product titres and/or the specific production rates (quantitative improvement). A further objective of great importance is the improvement of yields. For example, the costs of the carbon source in the case of penicillin production are very important and the fraction of the sugar which is converted into penicillin is only 10–15% [8].

Table 6. Criteria for strain improvement

Target	Impact on process or product
Improvement of titre and/or specific production rate	General decrease of production costs, improved exploitation of reactor capacity, lower investment costs, increased efficiency in downstream processing steps.
Improvement of yield	Lower costs for substrates, decreased production of heat and CO_2, lower cooling costs, less waste and pollution.
Changes in catabolic capabilities	Use of more favourable substrates (less expensive, better availability, etc.), omission of pretreatment steps (e.g. enzymatic hydrolysis of polysaccharides).
Improvement of technological features of microorganisms (e.g. flocculation behaviour, structure of mycelium, sporulation, foaming, strain stability, etc.)	Less energy costs for mixing and oxygen transfer, improved separation characteristics, fewer problems in inocculum preparation or scale-up of the process.
Improvement of product quality	Decreased production of specific by-products (fewer impurities), prevention of product degradation (e.g. proteinases).
Modification of products	Improvement of solubility in extraction solvents (e.g. addition of specific side chains), increased (thermic) stability of altered enzymatic properties of proteins.
Changing the locus of product accumulation (e.g. intracellular — extracellular)	Improved product recovery (e.g. omission of cell disruption), correct products (e.g. fully processed proteins).

However, additional criteria influencing the overall productivity of a bioprocess must be taken into account. In particular, such criteria can be orientated to improving each integrated process including both upstream and downstream steps. Further goals are connected with product quality and the efficiency of related processes such as effluent treatment and pollution control.

Some of these criteria have attained a special significance because of the possibility of applying rDNA techniques. Some important objectives for strain improvement in conjunction with the application of rDNA technology are listed in Table 6.

In general, living cells synthesize two different groups of products:

1) Direct gene products (DNA, RNA, protein). Synthesis is usually directed by the expression of a single gene.
2) Products synthesized via biochemical pathways which are controlled by a number of genes (primary and secondary metabolites, biopolymers, biomass, etc.)

Except for the small scale production of diagnostics and certain tools for molecular biology research (e.g. hybridization probes, cloning vectors and other specific DNA fragments), the group of direct gene products is mainly connected with the production of proteins. The benefit of rDNA technology in this field is obvious since any manipulation which enhances the expression of the one gene encoding for the desired protein will represent a general possibility for improving productivity.

Amplification and/or fusion of the respective gene to genetic elements which direct (regulated) strong expression (promoters, terminators, ribosome binding sites, upstream elements, etc.) or mediate efficient secretion (signal sequences) represent the most common strategies. In addition, modifications of any of these elements by specific in vitro manipulations can also be performed [18, 208].

However, as has already been discussed in Sect. 6, the whole complex of gene expression and its regulation, and in the case of secreted proteins in addition the intricated mechanisms of protein export must also be taken into account. Therefore, even the improvement of a process where the synthesis of the product is essentially only dependent on the expression of a single gene may involve difficult barriers, especially when heterologous genes are to be expressed.

The application of rDNA technology to the production of multigenic products is correspondingly more complex since many genes or their respective products (enzymes, regulatory proteins) are involved in the synthesis of a specific product. With respect to quantitative improvements (titres or specific production rates), two basic strategies can be applied:

1) Increasing the expression of genes coding for enzymes catalyzing rate-limiting biosynthetic steps.
2) Fusion of enzyme-coding genes which are subject to regulated expression (repression) to non-regulated expression systems which are no longer repressible.

8.2 Practical Examples

Applications of rDNA technology to solve strain improvement problems are now increasing exponentially. However, the situation is that most of these developments are being carried out by commercial institutions. Thus, a high level of actuality can hardly be attained by an academic scientist because of the high degree of secrecy existing in this field. The examples described here are therefore intended to show possibilities and trends rather than presenting a detailed overview of current activities.

8.2.1 Single Cell Protein and Commercial Enzymes

One of the first examples of the industrial use of genetically engineered organisms was connected with the production of single-cell protein. The energy-consuming pathway for the assimilation of ammonia in *Methylotrophus methylophilus* was replaced by an energy-conserving pathway based on *E. coli* glutamate dehydrogenase which resulted in a much improved yield [314].

In the field of the production of industrial enzymes, numerous reports on the cloning of the respective genes have been published. However, only few reports can be found referring to large-scale production with genetically engineered microorganisms. Recently, the production of α-amylase with a *Bacillus subtilis* strain containing a recombinant plasmid with the *Bacillus amyloliquefaciens* α-amylase gene was reported. On a laboratory scale (10 dm^3), the recombinant strain produced about twice the amount of α-amylase in comparison to a production strain which was developed by traditional methods [315]. The stability of the recombinant plasmid was sufficiently high to allow batch production of this enzyme [257].

An engineered, industrial strain of *Streptomyces violaceoniger* carrying multiple copies of a xylose isomerase gene in an integrated state has been used for the production of xylose isomerase in large tanks. A 3–5 fold increase in enzyme production was obtained. Furthermore, it was in this case not necessary to employ expensive xylose normally required as an inducer [316].

8.2.2 Primary and Secondary Metabolites

Considerable progress can be observed in the field of amino acid production. Various genes of the biosynthetic pathways for several amino acids have been cloned from several microbial strains. The strategies used till now for constructing overproducing strains are mostly based on increasing either the dosage of single genes or of a whole set of genes responsible for the respective biosynthetic pathway. The best documented examples are the overproduction of L-tryptophan [73, 306, 317], L-threonine [318–320], and L-lysine [321, 322] by *E. coli* strains. Recombinant overproducing strains of *Corynebacterium glutamicum* for the production of L-phenylalanine [323] and L-threonine [324], and of *Serratia marcescens* for the production of L-histidine, L-proline and L-threonine [325] have also been recently described.

In addition, the use of immobilized genetically engineered microorganisms which can highly express the biocatalyst required has been shown to be a powerful strategy for the production of L-serine and L-tryptophan [327] as well as glutathione [328] in enzyme bioreactor processes.

At the present time antibiotics still represent the most important sector in the biotechnology world market. As a consequence of this, much research has been done in order to apply rDNA techniques to antibiotic producing microoganisms. Although only very limited knowledge is presently available regarding the enzymology and regulation of most antibiotic biosynthetic pathways, some progress has already been achieved with streptomycetes. In addition to the development of vector systems, many genes involved in the production of several antibiotics have been identified and cloned [329–333]. Cloning of a fragment containing a gene involved in streptomycin production on a multicopy plasmid resulted in a seven-fold increase of the titre [333]

In the case of fungal antibiotic producers, the genes for the isopenicillin N synthetase (IPNS) of *Cephalosporium acremonium* [334] and of *Penicillium chrysogenum* [335] have been cloned. Since the last steps in the synthesis of β-lactam antibiotics are most probably the limiting reactions [336] these genes are of substantial interest. In a recent presentation of preliminary data it was reported that among several hundred transformants of *C. acremonium* carrying multiple integrated copies of the cloned IPNS gene, one strain with improved cephalosporin C titre was found [337].

Very interesting results could be obtained with respect to the construction of strains which synthesize modified antibiotics. By transferring cloned genes between *Streptomyces* strains producing actinorhodin, granaticin and medermycin, novel compounds representing "hybrid antibiotics" were found to be produced in recombinant strains [338, 339].

8.2.3 Qualitative Improvements

One most important target for strain improvement by rDNA techniques is the introduction of new catabolic capabilities into a production strain. For instance with "high volume — low price" products, substrate costs represent a high percentage of total production costs. Improvement of the ability to utilize specific substrates is, therefore, of great importance for such processes. Various examples of the introduction of genes mediating the utilization of specific substrates have already been described. Gene fragments coding for enzyme activities which hydrolyze sucrose have been cloned into *E. coli* strains permitting growth on sucrose-containing media (e.g. molasses)[340, 341]. Much effort has been put into cloning genes coding for amylolytic enzymes into *S. cerevisiae* strains. α-amylase genes from various sources and glucoamylase genes from filamentous fungi and yeasts have been successfully used for the construction of both laboratory and commercial starch-degrading strains of *S. cerevisiae*[108]. In addition, cellulase genes from *Trichoderma reesei*[342] and a xylose isomerase gene from *B. subtilis*[343] have been cloned and expressed in *S. cerevisiae*.

8.2.4 Applications in Food Industry and Agriculture

The principle microorganisms used for the production of food and beverages are yeasts and lactic acid bacteria. One particular goal of the brewing industry is the production of low carbohydrate beers[344]. Since the majority of *S. cerevisiae* strains have no ability to degrade dextrins, light beers are usually produced by enzymatic degradation of the dextrins by fungal glucoamylases. The cloning of a glucoamylase gene (*DEX1*) into a commercial brewing lager strain has already been reported[108]. Recently, the construction of a rDNA yeast strain capable of being used for the production of light beer has been claimed by a genetic engineering company[345].

A further target in beer production is the removal of water soluble β-glucans which can cause serious filtration problems. A gene fragment encoding a (1→3, 1→4)-β-D-glucanase activity was cloned into yeast. The gene was expressed and the amount of glucanase released in laboratory-scale experiments carried out under brewing conditions quantitatively degraded β-glucan added in concentrations known to be troublesome in beer production[346].

Considerable interest has been directed to the lactic acid bacteria used as starter cultures in the dairy industry. Since many of the technologically important functions are plasmid encoded, rDNA techniques are viewed as being potentially useful for the construction of improved strains[347]. It is expected that within a few years genetically engineered strains can be ready for use in commercial processes. Targets for improvement are lactate production, proteolysis, bacteriophage resistance and diacetyl production[348].

Nitrogen fixation is one of the main targets for agricultural applications of rDNA technology. In the USA, and in several other countries as well, there is already a small industry which provides preparations of specific *Rhizobium* strains for inoculation of leguminous crops to ensure the establishment of a nitrogen-fixing symbiosis[349]. Current improvement strategies are targeted at improving the efficiency of nitrogen fixation and the nodulation process. A major aim is to establish symbiotic nitrogen fixation systems with crops not naturally capable of symbiosis with nitrogen-fixing soil bacteria[350].

Production of biological pest control agents represents a further potential use of genetically engineered microorganisms. Various strains of *Bacillus thuringiensis* producing insecticidal glycoprotein crystals within the sporulating bacterial cell are presently used to produce toxin preparations[351]. Improvement of the toxin production by enhancing the expression of the respective toxin genes is one target for strain improvement [352]. In addition, the insertion of such endotoxin genes into root-colonizing bacteria such as *Pseudomonas fluorescens* has been performed and it is expected that such constructed strains can be applied as "living bioinsecticides" [353].

9 Concluding Remarks and Perspectives

The last few years have brought about considerable progress in the development of cloning systems with nearly all species of microorganisms of commercial interest. A good foundation is now available to go beyond the "high price — low volume" polypeptides of pharmaceutical interest and to apply rDNA technology to a broad spectrum of biotechnological processes.

Obviously, the class of products being manufactured by genetically engineered strains will in the first instance be industrial enzymes which in most cases represent products of single genes. Recent progress in the development of cloning systems for filamentous fungi which are the most important producers of commercial enzymes will certainly be a good basis for this.

With respect to the production of primary and secondary metabolites (synthesis controlled by many genes), many problems remain which must still be extensively studied. This especially applies to the development of systems which allow the controlled regulation of the expression of specific genes to a distinct level. There is not only a lack of information on the molecular biology of gene expression and its regulation, but very often there is even a lack of understanding of the biochemical and physiological background of the biosynthesis of metabolites, especially in the case of secondary metabolites.

rDNA techniques will not completely replace the "classical" methods of strain improvement, but offer additional powerful tools of increasing significance, Finally, it should not be forgotten that success in the application of genetic engineering strategies strongly depends on sufficient knowledge of the biology, genetics and physiology of the organisms used. The most sophisticated gene constructions will not be effective in an unsuitable cellular and/or external environment.

With respect to the impacts of rDNA technology on future developments two particular aspects should be mentioned here:

1) rDNA techniques allow the alteration of protein sequences at specific points. On the basis of increasing knowledge of the relationships between molecular structure and function of proteins and because of the rapid developments in computer sciences [354], the molecular design of protein molecules having desirable, specific catalytic or other functional properties is not longer an unfounded speculation. Protein engineering is already a rapidly developing field [355,356] and examples for altering the reaction characteristics, substrate specificities or thermal stabilities of enzymes have already been described for several systems [357,358].
2) Processes in the field of enzyme technology have been developed to an industrial scale [359]. Furthermore, several enzymes were found to be functional in organic solvents even at high temperatures [360–362]. These facts together with the

developments in protein engineering, may be the basis for a further increase in the applications of biotechnology in a widest sense.

10 Acknowledgement

I would like to thank Prof. R. M. Lafferty for critical reading of the manuscript. Some data presented here are from the project 5384 supported by the Austrian "Fonds zur Förderung der Wissenschaftlichen Forschung".

11 References

1. EFB definition of biotechnology, EFB Newsletter (1982) 5: 2
2. Drews SW, Wallis DA (1983) Regulation of secondary metabolism and keys to its manipulation, in: Bennett JW, Ciegler A (eds) Secondary metabolism and differentiation in fungi, Marcel Dekker, Inc. New York, p 35
3. Neijssel OM, Tempest DW (1985) Overproduction of microbial metabolites, in: Moo-Young M (ed) Comprehensive biotechnology, Vol 1, Pergamon Press, Oxford, p 605
4. Calam CT (1986) Physiology of the overproduction of secondary metabolites, in: Vanek Z, Hostalek Z (eds) Overproduction of microbial metabolites. Strain improvement and process control strategies. Butterworths, Boston, p 27
5. Fiechter A (1986) Bioprocess development, in: ibid., Butterworths, Boston, p 231
6. Lein J (1984) ASM News 49: 576
7. Demain AL (1966) Adv. Appl. Microbiol. 8: 1
8. Van der Beek CP, Roels JA (1984) A. v. Leeuwenhoek 50: 625
9. Van der Beek CP et al (1984) Progr. Ind. Microbiol. 20: 457
10. Lein J (1986) The Panlabs penicillin strain improvement program, in: Vanek Z, Hostalek Z (eds) Overproduction of microbial metabolites. Strain improvement and process control strategies. Butterworths, Boston, p 105
11. Vournakis JN, Elander RP (1983) Science 219: 703
12. Kikuchi M (1980) Biotech. Bioeng. 22 Supp. 1: 195
13. Demain AL (1972) Adv. Appl. Microbiol. 16: 177
14. Elander RP (1976) Mutation to increased product formation in antibiotic-producing microorganisms, in: Microbiology 1976, American Society for Microbiology, Washington D.C.
15. Elander RP, Vournakis JN (1986) Genetic aspects of overproduction of antibiotics and other secondary metabolites, in: Vanek Z, Hostalek Z (eds) Overproduction of microbial metabolites. Strain improvement and process control strategies. Butterworths, Boston, p 63
16. Nakayama K (1986) Progr. Ind. Microbiol. 24: 3
17. Rowlands RT (1984) Enzyme Microb. Technol. 6: 3
18. Rowlands RT (1984) ibid 6: 290
19. Ferenczy L et al (1974) Nature 248: 793
20. Fodor K, Ferenczy L (1976) Proc. Natl. Acad. Sci. USA 73: 2147
21. Schaeffer P et al (1976) ibid 73: 2151
22. Ferenczy L (1984) Fungal protoplast fusion: Basic and applied aspects, in: Beers Jr RF, Bassett EG (eds) Cell fusion: gene transfer and transformation, Raven Press, New York, p 145
23. Hopwood DA et al (1977) Nature 268: 171
24. Ferenczy L (1981) Microbial protoplast fusion, in: Glover WS, Hopwood DA (eds) Genetics as a tool in microbiology, Cambridge University Press, Cambridge, p 1
25. Peberdy JF (1987) Microbiol. Sciences 4: 108
26. Kurth R, Demain AL (1984) Drugs Pharm. Sci. 22: 781
27. Hamlyn PE, Ball C (1979) Recombination studies with *Cephalosporium acremonium*, in: Sebek OK, Laskin AI (eds) Genetics of industrial microbiology, American Society of Microbiology, Washington, DC, p 185
28. Wesseling AC (1982) Dev. Ind. Microbiol. 23: 31

29. Ferenczy L (1985) Protoplast fusion in yeasts, in: Peberdy JF, Ferenczy L (eds) Fungal protoplasts, Marcel Dekker, New York, p 279
30. Ball C (1985) Perspectives for protoplasts in the genetics of industrially important fungi, in: ibid. p 337
31. Hopwood DA (1981) Ann. Rev. Microbiol. 35: 237
32. Anne J, Peberdy JF (1985) Protoplast fusion and interspecies hybridization in Penicillium, in: Peberdy JF, Ferenczy L (eds) Fungal protoplasts, Marcel Dekker, New York, p 259
33. Kevei F, Peberdy JF (1985) Interspecies hybridization after protoplast fusion in Aspergillus, in: ibid. p 241
34. Peberdy JF (1983) Genetic recombination in fungi following protoplast fusion and transformation, in: Smith JE (ed) Fungal differentiation, Marcel Dekker, New York, p 559
35. Tamaki H (1986) Curr. Genet. 10: 491
36. Vane J, Cuatrecasas P (1984) Nature 312: 303
37. Riggs D et al (1984) From somatostatin to human insulin, in: Bollon AP (ed) Recombinant DNA Products: Insulin, interferon and growth hormone CRC Press, Boca Raton, p 38
38. Powledge TM (1984) Bio/Technology 2: 214
39. Holmes WE et al (1985) ibid. 3: 923
40. Westermann P (1986) BioEngineering 4: 32
41. Khosrovi B, Gray PP (1985) Products from recombinant DNA, in: Moo-Young M (ed) Comprehensive biotechnology, Vol 3, Pergamon Press, Oxford, p 319
42. Harris TJR, Emtage JS (1986) Microbiol. Sciences 3: 28
43. Doi RH et al (1986) Trends Biotechnol. 4: 232
44. Kingsman SM et al (1987) ibid. 5: 53
45. Maniatis T, Fritsch EF, Sambrook J (1982) Molecular cloning. Cold Spring Harbor NY, CSH Laboratory
46. Dahl HH et al (1981) The use of genomic libraries for the isolation and study of eukaryotic genes, in: Williamson R (ed) Genetic engineering, Vol 2, Academic Press, London p 49
47. Williams JG (1981) The preparation and screening of a cDNA clone bank, in: ibid. p 49
48. Gassen HG, Lang A (1982) Chemical and enzymatic synthesis of gene fragments. Verlag Chemie, Weinheim, FRG
49. Glover DM (1985) DNA Cloning, Vol 1, IRL Press, Oxford
50. Pritchard RH, Holland IB (1985) Basic cloning techniques, Blackwell, Oxford
51. Hindley J (1983) DNA Sequencing, Elsevier, Amsterdam
52. Hames BD, Higgins SJ (1985) Nucleic acid hybridization, IRL Press, Oxford
53. Thomasz A (1970) Ann. Rev. Genet. 3: 217
54. Cohen SN et al (1972) Proc. Natl. Acad. Sci. USA 69: 3240
55. Chakrabarthy AM et al (1975) ibid. 72: 3647
56. Sano A, Kagevama M (1977) J. Gen. Appl. Microbiol. 23: 183
57. Schwab H (1981) Genetic Engineering, in: Lafferty RM (ed) Fermentation, Springer, Wien, p 212
58. Selvaraj G, Iyer VN (1981) Gene 15: 279
59. David M et al (1981) J. Bacteriol. 146: 1154
60. Hardy KG (1985) *Bacillus* cloning methods, in: Glover DM (ed) DNA cloning, Vol 2, IRL Press, Oxford, p 1
61. Hunter IS (1985) Gene Cloning in *Streptomyces*, in: ibid, p 19
62. Rothstein R (1985) Cloning in Yeast, in: ibid, p 45
63. Case ME (1983) Nato Adv. Study Inst. Ser., Ser. A 61: 1
64. Turner G, Balance DJ (1985) Cloning and transformation in *Aspergillus*, in: Bennett JW, Lasure LL (eds) Gene manipulation in fungi, Academic Press, Orlando, p 259
65. Hamlyn PF et al (1981) Enzyme Microb. Technol. 3: 321
66. Shimizu-Kadota K, Kudo S (1984) Agric. Biol. Chem. 48: 1105
67. Makins JF, Holt G (1981) Nature 293: 671
68. Radford A et al (1981) Mol. Gen. Genet. 184: 567
69. Ito H et al (1983) J. Bacteriol. 153: 163
70. Dhawle SS et al (1984) Curr. Genet. 8: 77
71. Zimmermann O (1986) Rev. Physiol. Biochem. Pharmacol. 105: 175
72. Hashimoto H et al (1985) Appl. Microbiol. Biotechnol. 21: 336

73. Imanaka T (1986) Adv. Biochem. Eng. Biotechnol. 33: 1
74. Datta N, Hedges RW (1972) J. Gen. Microbiol. 70: 453
75. Simon R et al (1982) Bio/Technology 1: 784
76. Guiney DG, Yakobson E (1983) Proc. Natl. Acad. Sci. USA 80: 3595
77. Krek W et al (1987) Replikation von RP4-Derivaten in *Alcaligenes eutrophus*, in: Biotechnologie und Gentechnik, Proc, Bundesministerium für Wissenschaft und Forschung, Wien (Vienna) p 96
78. Priefer U et al (1985) J. Bacteriol. 163: 324
79. Ditta G et al (1980) Proc. Natl. Acad. Sci. USA 77: 7347
80. Boulnois F (1981) Mol. Gen. Genet. 182: 152
81. Lydiate DJ et al (1985) Gene 35: 223
82. Kieser T (1984) Riv. Biol. 77: 551
83. Oultram JD, Young M (1985) FEMS Microbiol. Lett. 27: 129
84. Cove DJ (1971) Heredity 36: 191
85. van Hartingsveldt W et al (1987) Mol. Gen. Genet. 206: 71
86. Buxton FP, Radford A (1983) ibid 190: 403
87. Kelly JM, Hynes M (1985) The EMBO J. 4: 475
88. Wolk CP et al (1984) Proc. Natl. Acad. Sci. USA 81: 1561
89. Franklin FHC (1985) Broad host range cloning vectors for Gram negative bacteria, in: Glover DM (ed) DNA Cloning, Vol 1, IRL Press, Oxford, p 1165
90. Roychoudhury R, Lam A (1983) Gene 24: 53
91. Rawlings DE et al (1984) J. Biotechnol. 1: 129
92. Ditta G et al (1985) Plasmid 13: 149
93. Yoshimura K et al (1983) Gene 24: 255
94. Brückner R et al (1984) ibid 32: 151
95. Thompson CJ et al (1982) ibid 20: 51
96. Yoshihama M et al (1985) ibid 162: 591
97. Katsumata R et al (1984) J. Bacteriol. 159: 306
98. Ozaki A et al (1984) Mol. Gen. Genet. 196: 175
99. Yeh P et al (1986) Gene 47: 301
100. Shrago AW et al (1986) Appl. Environm. Microbiol. 52: 574
101. Macrina FL et al (1983) Gene 25: 145
102. Squires CH et al (1984) J. Bacteriol. 159: 465
103. Beggs JD (1981) Gene cloning in yeast, in: Williamson R (ed) Genetic engineering, Vol 2, Academic Press, London, p 175
104. Webster TD, Dickson RC (1983) Gene 26: 243
105. Kaster KR et al (1984) Curr. Genet. 8: 353
106. Zhu J et al (1985) Bio/Technology 3: 451
107. Meacock PA et al (1986) A dominant marker for plasmid transformation of commercial strains of yeast, in: Fifth international symposium on the genetics of industrial microorganisms, book of abstracts, Split
108. Tubb RS (1986) Trends Biotechnol. 4: 98
109. Cegg JM et al (1985) Mol. Cell. Biol. 5: 3376
110. Kawamura M et al (1983) Gene 24: 157
111. Beach D, Nurse P (1981) Nature 290: 140
112. Das S et al (1984) J. Bacteriol. 158: 1165
113. Nilsson-Tillgren T et al (1984) Riv. Biol. 77: 609
114. Nellen W, Fiertel RA (1985) Gene 39: 155
115. Van Heeswijck R, Roncero MI (1984) Carlsberg Res. Commun. 49: 691
116. Revuelty JL, Jayaram M (1986) Proc. Natl. Acad. Sci. USA 83: 7344
117. Case ME et al (1979) ibid. 76: 5259
118. Schechtman MG, Yanofsky C (1983) J. Mol. Appl. Genet. 2: 83
119. Yelton MM et al (1984) Proc. Natl. Acad. Sci. USA 81: 1470
120. Begueret J et al (1984) Gene 32: 487
121. John MA, Peberdy JF (1984) Enzyme Microb. Technol. 6: 386
122. Buxton FP et al (1985) Gene 37: 207
123. Cantoral JM et al. (1987) Bio/Technology 5, 323

124. Iimura Y et al (1987) Agric. Biol. Chem. 51: 323
125. Queener SW et al (1985) A system for genetic transformation of *Cephalosporium acremonium*, in: Leive L (ed) Microbiology 1985, American Society for Microbiology, Washington DC, p 468
126. Penalva MA et al (1985) Studies on transformation of *Cephalosporium acremonium*, in: Timberlake WE (ed) Molecular genetics of filamentous fungi, Alan R. Liss. Inc., New York, p 59
127. Kolar M et al (1988) Gene 62: 127
128. Turner G, Ward M (1986) An antibiotic resistance marker for fungal transformation, in: Bailey J (ed) NATO ASI Series, Vol H1, Biology and molecular biology of plant-pathogen interaction, Springer, Berlin, p 359
129. Orbach MJ et al (1986) Mol. Cell. Biol. 6: 2452
130. Munoz-Rivas A et al (1986) Mol. Gen. Genet. 205: 103
131. Rochaix J-D, van Dillewijn J (1982) Nature 296: 70
132. Balbás P et al (1986) Gene 50: 3
133. Kaiser K, Murray NE (1985) The use of phage Lambda replacement vectors in the construction of representative genomic DNA libraries, in: Glover DM (ed) DNA Cloning, Vol 1, IRL Press, Oxford, p 1
134. Huynh TV et al (1985) Construction and screening of cDNA libraries in λgt10 and λgt11, in: ibid, p 49
135. Messing J (1981) M13mp2 and derivatives: A molecular cloning system for DNA sequencing, strand specific hybridization and in vitro mutagenesis, in: Walton AG (ed) Recombinant DNA, Elsevier, Amsterdam, p 143
136. Collins J, Hohn B (1978) Proc. Natl. Acad. Sci. USA 75: 4242
137. Grosveld FG, Dahl HHM (1983) Cosmid libraries, in: Techniques in the life sciences, B5, Nucleic Acid Biochemistry, B502, Elsevier, Ireland, p 1
138. Pouwels PH et al (1985) Cloning vectors. Elsevier, Amsterdam
139. Figurski DH et al (1985) Basic Life Sci. 30: 227
140. Schwab H et al (1983) Arch. Microbiol. 136: 140
141. Werneke JM et al (1985) Gene 38: 73
142. Davison J et al (1987) ibid. 51: 275
143. Leemans R et al (1987) J. Bacteriol. 169: 1899
144. Leemans J et al (1982) Gene 19: 361
145. Tait RC et al (1983) Bio/Technology 1: 269
146. Kieselburg MK et al (1984) ibid 2: 254
147. Imanaka T (1983) Trends Biotechnol. 1: 139
148. Ishiwa H, Tsuchida N (1984) Gene 32: 129
149. Sullivan MA et al (1984) ibid 29: 21
150. Ulmanen I et al (1985) J. Bacteriol. 162: 176
151. Prozorov AA et al (1985) Gene 34: 39
152. Fayerman JF et al (1985) Development of systems for heterologous gene expression in *Streptomyces* spp., in: Leive L (ed) Microbiology 1985, American Society for Microbiology, Washington DC, p 414
153. Wohlleben W, Pühler A (1984) Development of shuttle vectors between *Streptomyces* species and *Escherichia coli*, in: Third European Congress on Biotechnology. Proc. Vol III, Verlag Chemie, Weinheim, p 111–219
154. Larson JL, Hershberger CL (1984) J. Bacteriol. 157: 314
155. Martin JF et al (1987) Bio/Technology 5: 137
156. Santamaria R et al (1984) J. Gen. Microbiol. 130: 2237
157. Santamaria R et al (1985) J. Bacteriol. 162: 463
158. Takagi H et al (1986) Agric. Biol. Chem. 50: 2597
160. Beppu T (1986) Progr. Ind. Microbiol. 24: 24
161. Miwa K et al (1985) Gene 39: 281
162. Kok J et al (1984) Appl. Environm. Microbiol. 48: 726
163. Macrina FL et al (1982) Molecular cloning strategies for the Streptococcus sanguis host vector system, in: Schlessinger D (ed) Microbiology 1982, American Society for Microbiology, Washington DC, p 234

164. Yu P-L, Pearce LE (1986) Biotechnol. Lett. 8: 469
165. Smith MD (1985) J. Bacteriol. 162: 92
166. Murray AW, Szostak JW (1983) Nature 305: 189
167. Blackburn EH (1985) Trends Genet. 2: 8
168. Hieter P et al (1985) Cell 40: 381
169. Zakian VA et al (1986) Mol. Cell. Biol. 6: 925
170. Crosby WL, Thomas DY (1983) Gene cloning in yeast, in: Techniques in the life sciences, B5, Nucleic Acid Biochemistry. B503, Elsevier, Ireland, p 1
171. Das S, Hollenberg CP (1982) Curr. Genet. 6: 123
172. de Louencourt L et al (1983) J. Bacteriol. 154: 737
173. Kurty MB et al (1986) Mol. Cell. Biol. 6: 142
174. Holmberg S (1984) Trends Biotechnol. 2: 98
175. Esser K et al (1986) Plasmids of eukaryotes, Fundamentals and applications. Springer, Berlin
176. Stohl LL, Lambowitz AM (1983) Proc. Natl. Acad. Sci. USA 80: 1058
177. Stahl U et al (1982) ibid 79: 3645
178. Skatrud PL, Queener SW (1984) Curr. Genet. 8: 155
179. Schwab H et al (1985) J. Cell. Biochem., Suppl. 9C 173
180. Ballance DJ, Turner G (1985) Gene 36: 321
181. van Heeswijk (1986) Carlsberg Res. Commun. 51: 433
182. Perrot M et al (1987) Mol. Cell. Biol. 7: 1725
183. Esser K, Mohr G (1986) Process Biochem 21: 153
184. Cullen D, Leong S (1986) Trends Biotechnol. 4: 285
185. Wernars K et al (1986) Curr. Genet. 9: 361
186. Tilburn J et al (1983) Gene 26: 205
187. Turner G et al (1985) Development of cloning vectors and a marker for gene replacement techniques in *Aspergillus nidulans*, in: Timberlake WE (ed) Molecular genetics of filamentous fungi, Alan R. Liss Inc., New York, p 15
188. Yelton MM et al (1985) Proc. Natl. Acad. Sci. USA 82: 834
189. Turgeon BG et al (1985) Mol. Gen. Genet. 201: 450
190. Kuiper MTR, de Vries H (1985) Curr. Genet. 9: 471
191. Akins RA, Lambowitz AM (1985) Mol. Cell. Biol. 9: 2273
192. Grant DM et al (1984) ibid 4: 2041
193. Sanchez F et al (1987) Gene 51: 97
194. Beri RK, Turner G (1987) Curr. Genet. 11: 639
195. Brygoo Y, Debuchy R (1985) Mol. Gen. Genet. 200: 128
196. Binninger DM et al (1987) EMBO J. 6: 835
197. Banks GR (1983) Curr. Genet. 7: 73
198. Rochaix J-D et al (1984) Cell. 36: 925
199. Mc. Clure WR (1985) Ann. Rev. Biochem. 54: 171
200. Reznikoff WS et al (1985) Ann. Rev. Genet. 19: 355
201. Reznikoff WS, Mc.Clure WR (1986) *E. coli* promoters, in: Reznikoff W, Gold S (eds) Maximizing gene expression, Butterworths, Boston, p 1
202. Holmes WM et al (1983) Cell 32: 1029
203. Stormo GD (1986) Translation initiation, in: Reznikoff W, Gold S (eds) Maximizing gene expression, Butterworths, Boston, p 195
204. Deuschle V et al (1986) EMBO J. 5: 2987
205. Kammerer W et al (1986) ibid 5: 2995
206. Nishi T, Itoh S (1986) Gene 44: 29
207. Chen CY et al (1984) Nucl. Acids Res. 23: 8951
208. Davies J (1987) Retrospect on heterologous gene expression, in: Alačević M, Hranneli D, Toman Z (eds) Genetics of industrial microorganisms, Proc. of the fifth international symposium, part A, Zagreb, p 17
209. McConnel DJ et al (1986) Ann. N.Y. Acad. Sci. 469: 1
210. Janssen GR et al (1985) Isolation and analysis of *Streptomyces* promoters, in: Leive L (ed) Microbiology 1985, American Society for Microbiology, Washington DC, p 392
211. Buttner MJ, Brown NL (1987) Gene 51: 179

212. Jaurin B, Cohen SN (1985) ibid 39: 191
213. Bibb MJ, Cohen SN (1982) Mol. Gen. Genet. 187: 265
214. Lovett PS, Schoner RG (1983) Gene Amplif. Anal. 3: 117
215. Jackson DA (1986) Trends Biochem. Sci. 11: 249
216. Guarente L (1984) Cell 36: 799
217. Struhl K (1986) Yeast promoters, in: Reznikoff W, Gold L (eds) Maximizing gene expression, Butterworths, Boston, p 35
218. McNeil JB, Smith M (1986) J. Mol. Biol. 187: 363
219. Hahn S et al (1986) Proc. Natl. Acad. Sci. USA 82: 8562
220. Braun RJ et al (1986) EMBO J. 5: 603
221. Guarente L et al (1984) Cell 36: 503
222. Tajima M et al (1986) Mol. Cell. Biol. 6: 246
223. Struhl K (1985) Proc. Natl. Acad. Sci. USA 82: 8419
224. Clements JM, Roberts C'· (1986) Gene 44: 97
225. Gwynne DI et al (1987) ibid 51: 205
226. Newbury SF et al (1986) ibid 43: 51
227. Goldberger RF (1979) Strategies of genetic regulation in procaryotes, in: Goldberger RF (ed) Biological regulation and development, Vol 1: Gene expression, Plenum Press, New York, p 1
228. Kolter R, Yanofsky C (1982) Ann. Rev. Genet. 16: 113
229. Pines O, Inoye M (1986) Trends Genet. 2: 284
230. Campbell KM et al (1983) Protein-mediated translational repression, in: Beckwith J, Davies J, Gallant JA (eds) Gene function in procaryotes, Cold Spring Harbor, CSH Laboratory, p 185
231. Buell G, Panayotatos N (1983) Mechanism and practice, in: Beckwith J, Davies J, Gallant JA (eds) Gene function in procaryotes, Cold Spring Harbor, CSH Laboratory, p 345
232. Stark GR, Wahl GM (1984) Ann. Rev. Biochem. 53: 447
233. Sninsky JJ et al (1981) Gene 16: 275
234. Larsen J et al (1984) ibid. 28: 45
235. van Brunt J (1986) Bio/Technology 4: 1057
236. Arst HN, Scazzocchio C (1985) Formal genetics and molecular biology of gene expression in *Aspergillus nidulans*, in: Bennett JW, Lasure LL (eds) Gene manipulations in fungi, Academic Press, Orlando, p 309
237. Huiet L, Case M (1985) Molecular biology of the *qa* gene cluster of *Neurospora*, in: ibid, p 229
238. Lockington RA et al (1985) Gene 33: 137
239. van Gorcom RFM et al (1986) ibid 48: 211
240. Palva I (1987) Protein secretion, in: Alačević M, Hranueli D, Toman Z (eds) Genetics of industrial microorganisms, Proc. of the fifth international symposium, part B, Zagreb, p 415
241. Holland IB et al (1986) Bio/Technology 4: 427
242. Nicaud J-M et al (1986) J. Biotechnol. 3: 255
243. Carter BLA et al (1986) Microbiol. Sci. 3: 23
244. Palva I et al (1987) Secretion of proteins in bacilli, in: Alačević M, Hranueli D, Toman Z (eds) Genetics of industrial microorganisms, Proc. of the fifth international symposium, part B, Zagreb, p 421
245. Innis MA et al (1985) Science 228: 21
246. Cullen D et al (1987) Bio/Technology 5: 369
247. Priest FG (1985) Extracellular enzymes, in: Moo-Young M (ed) Comprehensive Biotechnology, Vol 1, Pergamon Press, Oxford, p 587
248. Zabriskie DW, Arcuri EJ (1986) Enzyme Microb. Technol. 8: 706
249. Williamson DH (1987) Genomic instability in microorganisms, in: Alačević M, Hranueli D, Toman Z (eds) Genetics of industrial microorganisms, Proc. of the fifth international symposium, part A, Zagreb, p 217
250. Reusser F (1963) Adv. Appl. Microbiol. 5: 189
251. Johnston JR (1975) Strain improvement and stability in filamentous fungi, in: Smith JE, Berry DR (eds) The filamentous fungi, Vol 1, Industrial mycology, Edward Arnold Publ. Ltd., London, p 59

252. Cesareni G et al (1987) *E. coli* plasmid stability and control of copy number, in: Alačevič M, Hranueli D, Toman Z (eds) Genetics of industrial microorganisms, Proc. of the fifth international symposium, part A, Zagreb, p 247
253. Primrose SB et al (1983) Spec. Publ. Soc. Gen. Microb. 10: 63
254. Jones SA, Melling J (1984) FEMS Microbiol. Lett. 22: 239
255. Rhodes PM et al (1987) Plasmid instability in *E. coli* expressing recombinant proteins, in: Alačević M, Hranueli D, Toman Z (eds) Genetics of industrial microorganisms, Proc. of the fifth international symposium, part A, Zagreb, p 259
256. Mizukami T et al (1986) Agric. Biol. Chem. 50: 1019
257. Vehmaanperä JO, Korhola MP (1986) Appl. Microbiol. Biotechnol. 23: 456
258. Summers DK, Sherratt DJ (1984) Cell 36: 1097
259. Gerdes K et al (1985) J. Bacteriol. 161: 292
260. Mori H et al (1986) J. Mol. Biol. 192: 1
261. Abeles AL et al (1985) ibid 185: 261
262. Miki T et al (1980) J. Bacteriol. 141: 87
263. Tucker WT et al (1984) Cell 38: 191
264. Saurugger PN et al (1986) J. Biotechnol. 4: 333
265. Hakkaart MJJ et al (1984) Cell 36: 203
266. Murray JAH, Cesareni G (1986) EMBO J. 5: 3391
267. Gustafsson P et al (1983) ibid 2: 27
268. Austin S et al (1986) J. Bacteriol. 168: 1010
269. Gerdes K et al (1986) EMBO J. 5: 2023
270. Hiraga S. et al (1986) J. Bacteriol. 166: 100
271. Skogman G et al (1983) Gene 23: 105
272. Yukawa H et al (1985) Agric. Biol. Chem. 49: 3619
273. Nilsson J, Skogman SG (1986) Bio/Technology 4: 901
274. Ferrari E et al (1985) ibid 3: 1003
275. Marquet M et al (1986) Biotechnol. Lett. 8: 535
276. Lauffenburger DA (1987) Trends Biotechnol. 5: 87
277. Rostock PR, Jr, Hershberger CL (1983) Gene 25: 29
278. Koshland D et al (1985) Cell 40: 393
279. Nugent ME et al (1983) Dev. Ind. Microbiol. 24: 271
280. Carrier MJ et al (1983) Trends Biotechnol. 1: 109
281. Rood JI et al (1980) J. Bacteriol. 144: 552
282. Starlinger P (1980) Plasmid 3: 241
283. Vögtli M et al (1986) Swiss Biotech. 4: 35
284. Hintermann G, Hütter R (1984) Riv. Biol. 77: 548
285. Schrempf H (1985) Genetic instability: amplification, deletion, and rearrangements within *Streptomyces* DNA, in: Leive L (ed) Microbiology 1985, American Society for Microbiology, Washington DC, p 436
286. Schrempf H et al (1987) Amplification and deletion of DNA sequences in *Streptomyces*, in: Alačević M, Hranueli D, Toman Z (eds) Genetics of industrial microorganisms, Proc. of the fifth international symposium, part A, Zagreb, p 177
287. Lopez P et al (1984) Proc. Natl. Acad. Sci. USA 81: 5189
288. Hahn J, Dubnau D (1985) J. Bacteriol. 162: 1014
289. Ehrlich SD et al (1987) Structural instability of genes cloned in *Bacillus subtilis*, in: Alačević M, Hranueli D, Toman Z (eds) Genetics of industrial microorganisms, Proc. of the fifth international symposium, part B, Zagreb, p 93
290. Upshall A (1986) Curr. Genet. 10: 593
291. Wernars K et al (1986) Mol. Gen. Genet. 205: 312
292. Durrens P et al (1986) ibid. 203: 544
293. Wouters JTM et al (1980) A. van Leeuwenhoek 46: 353
294. Walmsley P et al (1983) Mol. Gen. Genet. 192: 361
295. Pinches A et al (1985) Biotechnol. Lett. 7: 621
296. Watson TG et al (1986) ibid 8: 687
297. Kleinman MJ et al (1986) ibid 8: 225
298. DeBernardez ER, Dhurjati PS (1987) Biotech. Bioeng. 29: 558

299. Imanaka T, Aiba S (1981) Ann. N.Y. Acad. Sci. 369: 1
300. Ollis DF (1982) Phil. Trans. R. Soc. Lond. B. 297: 617
301. Ollis DF, Chang H-T (1982) Biotech. Bioeng. 24: 2583
302. Bailey JE et al (1983) Ann. N.Y. Acad. Sci. 413: 71
303. Bailey JE et al (1986) ibid 469: 194
304. Sardonini SA, DiBiasio D (1987) Biotech. Bioeng. 29: 469
305. Ryu DDJ, Siegel R (1986) Ann. N.Y. Acad. Sci. 469: 73
306. Imanaka T (1983) Biotech. Advs. 1: 279
307. Ryder DF, DiBiasio D (1984) Biotech. Bioeng. 26: 942
308. Seo J-H, Bailey JE (1986) ibid 28: 1590
309. Seressiotis A, Bailey JE (1987) ibid 29: 392
310. Shimizu N et al (1987) J. Ferment. Technol. 65: 7
311. Bailey FJ et al (1987) J. Ind. Microbiol. 2: 47
312. Bull DN (1983) Bio/Technology 1: 847
313. Hopkins DJ et al (1987) Biotech. Bioeng. 29: 85
314. Dart EC et al (1981) Genetic Engineering, in: Norris JR, Richmond MH (eds) Essays in Applied Microbiology, Wiley & Sons Ltd., London, p 3/1
315. Vehmaanperä J et al (1986) Industrial production of α-amylase by genetically engineered *Bacillus*, in: Fifth international symposium on the genetics of industrial microorganisms, book of abstracts, Split, p 36
316. Tiraby G et al (1986) Industrial production of glucose isomerase by a genetically engineered strain of *Streptomyces violaceoniger*, in: ibid, p 36
317. Aiba S et al (1982) Appl. Environm. Microbiol. 43: 289
318. Miwa K et al (1983) Agric. Biol. Chem. 47: 2329
319. Miwa K et al (1984) ibid 48: 2233
320. Debabov V (1983) Construction of strains producing L-threonine, in: Ikeda Y, Beppu T (eds) Genetics of industrial microorganisms 1982, Proc. of the fourth international symposium Kodansha Ltd., Tokyo, p 254
321. Dauce-LeReverend B et al (1982) Europ. J. Appl. Microbiol. Biotechnol. 15: 227
322. Tosaka O et al (1983) Trends Biotechnol. 3: 70
323. Ozaki A et al (1985) Agric. Biol. Chem. 49: 2925
324. Katsumata R et al (1987) Threonine production by the lysine producing strain of *Corynebacterium glutamicum* with amplified threonine biosynthetic operon, in: Alačević M, Hranueli D, Toman Z (eds) Genetics of industrial microorganisms, Proc. of the fifth international symposium, part B, Zagreb, p 217
325. Kisumi M (1987) Construction of amino acid-hyperproducing strains of *Serratia marcescens* by transductional and recombinant DNA techniques, in: ibid, p 253
326. Niederberger P et al (1987) Genetic manipulation of tryptophan biosynthesis in yeast, in: ibid, p 263
327. Hamilton BK et al (1985) Trends Biotechnol. 3: 64
328. Kimura A (1986) Adv. Biochem. Eng. Biotechnol. 33: 29
329. Fayerman JT (1986) Bio/Technology 4: 786
330. Malpartida F, Hopwood DA (1984) Nature 309: 462
331. Martin JF, Gil JA (1984) Bio/Technology 2: 63
332. Stanzak R et al (1986) ibid 4: 229
333. Kumada Y et al (1986) Gene 42: 221
334. Samson SM et al (1985) Nature 318: 191
335. Carr LG et al (1986) Gene 48: 257
336. Martin JF, Livas P (1985) Trends Biotechnol. 3: 39
337. Skatrud PL et al (1987) Improved transformation of *Cephalosporium acremonium*, in: Alačević M, Hranueli D, Toman Z (eds) Genetics of industrial microorganisms, Proc. of the fifth international symposium, part B, Zagreb, p 111
338. Hopwood DA et al (1985) Nature 314: 642
339. Floss HG (1987) Trends Biotechnol. 5: 111
340. Garcia JL (1985) Mol. Gen. Genet. 201: 575
341. Friehs K et al (1986) J. Biotechnol. 3: 333
342. Knowles JKC et al (1985) J. Cell. Biochem. Suppl. 9C 188

343. Wilhelm W et al (1984) Riv. Biol. 77: 607
344. Stewart GG (1985) J. Am. Soc. Brew. Chem. 43: 61
345. Biotechnol. News (1986) 6: Nov 26, 4
346. von Wettstein D (1987) Molecular genetics in the improvement of brewer's and distiller's yeast, in: Alačević M, Hranueli D, Toman Z (eds) Genetics of industrial microorganisms, Proc. of the fifth international symposium, part A, Zagreb, p 107
347. Teuber M (1986) Swiss Biotech. 4: 16
348. Venema G, Kok J (1987) Trends Biotechnol. 5: 144
349. Stacey G, Upchurch RG (1984) ibid 2: 65
350. Sprent JI (1986) ibid 4: 124
351. Khachatourians GG (1986) ibid 4: 120
352. Kristiansen B, Lewis C (1986) ibid 4: 56
353. Obukowicz MG et al (1987) Gene 51: 91
354. Wilson T, Klausner A (1984) Bio/Technology 2: 511
355. Leatherbarrow RJ, Fersht AR (1986) Protein Engineering 1: 7
356. Hol WGJ (1987) Trends Biotechnol. 5: 137
357. Wilkinson AJ et al (1984) Nature 307: 187
358. Querol E, Parrilla A (1987) Enzyme Microb. Technol. 9: 238
359. Weetall HH, Pitcher WH, Jr (1986) Science 232: 1396
360. Carrera G (1984) Trends Biotechnol. 2: 102
361. Grüninger H (1984) Swiss Biotech. 2: 35
362. Zaks A, Klibanov AM (1985) Proc. Natl. Acad. Sci. USA 82: 3192
363. Punt PJ et al (1987) Gene 52: 117
364. Skatrud PL et al (1987) Curr. Genet. 12: 337
365. Pentillä M et al (1987) Gene 61: 155